D1064324

Executing the Supply Chain

Executing the Supply Chain

Modeling Best-in-Class Processes and Performance Indicators

Alexandre Oliveira
Anne Gimeno

Associate Publisher: Amy Neidlinger
Executive Editor: Jeanne Glasser Levine
Operations Specialist: Jodi Kemper
Cover Designer: Chuti Prasertsith
Managing Editor: Kristy Hart
Project Editor: Andy Beaster
Copy Editor: Keith Cline
Proofreader: Sarah Kearns
Indexer: Lisa Stumpf
Compositor: Nonie Ratcliff
Manufacturing Buyer: Dan Uhrig

Upper Saddle River, New Jersey 07458

For information about buying this title in bulk quantities, or for special sales opportunities (which may include electronic versions; custom cover designs; and content particular to your business, training goals, marketing focus, or branding interests), please contact our corporate sales department at corpsales@pearsoned.com or (800) 382-3419.

For government sales inquiries, please contact governmentsales@pearsoned.com.

For questions about sales outside the U.S., please contact international@pearsoned.com.

Printed in the United States of America

First Printing July 2014

ISBN-10: 0-13-376438-9
ISBN-13: 978-0-13-376438-3

Pearson Education LTD.
Pearson Education Australia PTY, Limited.
Pearson Education Singapore, Pte. Ltd.
Pearson Education Asia, Ltd.
Pearson Education Canada, Ltd.
Pearson Educación de Mexico, S.A. de C.V.
Pearson Education—Japan
Pearson Education Malaysia, Pte. Ltd.

Library of Congress Control Number: 2014937235

Contents

Acknowledgments

As mentioned in the Authors' Note, we are currently writing five books for Pearson, all of which benefit from contributions by a large number of individuals from both academic and business sectors.

Our special thanks to Professor Martin Christopher, Emeritus Professor of Marketing & Logistics at Cranfield School of Management, Cranfield University, United Kingdom, whose contributions are present in all five of our books. We also wish to express our kindest appreciation to Mark Barratt, Associate Professor in Supply Chain Management at Marquette University, USA, and expert in collaboration and visibility across supply chains, for his contribution during the early stages of our academic background.

We would also like to thank Professor Milton Mori (State University of Campinas, Unicamp, Brazil), Professor Mauro Sampaio (University Center, FEI, Brazil), Professor Richard Wilding OBE (Cranfield University, UK), Lecturer Carla Corte (State University of Campinas, Unicamp, Brazil), Professor Armando Dal Colleto (Business School São Paulo, BSP, Brazil), and Senior Lecturer Melvin Peters (Cranfield University, UK) for their contributions to our development over the last 20 years.

In our books, business cases/expert opinions were provided by the following:

Bernardo Faria: Operations Development at Cielo, Brazil; Cesar Righetti, Operations Director at Cielo, Brazil; Dr Alan Smart, Senior Lecturer at Cranfield School of Management, UK; Enrique Motilla, founding partner at Quad Tree S.A. de C.V., Mexico; Fikri Dweiri, Associate Professor in IndusOperation Engineering and Management Department and Vice Dean of College of Engineering at the University of Sharjah (UoS), UAE; Gianluca S. Cesare, Director at

Lombardia Informatica, Italy; Kimmochi Eguchi, Executive Director, International Marketing Institute, Japan; Lucas Costa: Operations Planning at Cielo, Brazil; Luciano Bortoncello, Development Director at Transmiro, Brazil; Marcelo Pereira, Transport Manager at Transmiro, Brazil; Mauricio Ajzenberg, General Director at SPDL, Brazil; Robin Parsons, Supply Chain Manager, Maxinutrition (a GSK company), UK; Sharfuddin Ahmed Khan, lecturer in IndusOperation Engineering & Management Department at University of Sharjah (UoS), UAE; Shogo Kurokawa, President, Nippon Logistech Corporation, Japan.

This project would have proven impossible without the inspiration from our colleagues active in several professional networks. The governance team of the Brazilian Institute of Supply Chain Professionals (www.ibpsc.net/IBS), together with all associates, creates a unique technical environment that keeps us up-to-date on the best practices. Although we want to thank all IBS members, it is impossible to list anything more here than the advisory board:

Daniel Mello, Daines Toledo, Marcelo Alencar, Maricea dos Santos, Sérgio Romero, Daniel Okino, Ralph Martins, Cláudio de Sá, Alex Rocha, Luis Gonzaga, Marcelo Torres, William Marques, Antonio Berna, Carlos Pavanelli, Eduardo Junqueira, Antonio Souza, Luciano Bortoncello, Gelber Abe, Fabio Miranda, Luis Silva, Carlos Cirillo, Plínio Márcio, Luiz Ribeiro, Daniel Hermeto, Luiz Ferreira, Marco Palmeira.

We also want to thank our colleagues at CEBRALOG who have collaborated to develop and organize a significant part of the knowledge shared in our books: Andre Moraes, Eraldo Bertagnoli, Fernanda Silva, Adriana Freitas, Mauricio Cortes, Mirella Gomes, Francismar Lemos, Lucas Casagrande, and Fernando Fedato.

The Operations and Supply Chain Academic Group at LinkedIn (with nearly 20,575 members) was another important source of inspiration.

My kindest thanks to my editor, Jeanne Levine, and to Barry Render, who believed in and supported this book project from the early stages, and to Heather Simpkins, my proofreader, without whom this work would not be readable!

Finally, we wish to thank our relatives who have been a constant source of wisdom: our grandparents Pedro, Antônia, Ouriques, Claudina, David, Conchita, Leopoldo, and Clara; our parents Gerusa, Jorge, Lluis, and Maria Helena; our brothers Alan, Amanda, Andre, and Arnau; our sons Gabriel and Pedro; our nieces Mariana and Maria Paula; and our dearest Ana Claudia, Cristina, Maria Eduarda, Maria Etienne, and Rayder.

About the Authors

Alexandre Oliveira is a founding-partner at CEBRALOG (www.cebralog.com), a supply chain consultancy and training company headquartered in Brazil since 2001. Alexandre has also been President of the Brazilian Institute of Supply Chain Professionals (I.B.S.) since 2007. He offers courses for MBA programs at Business School São Paulo and is a regular contributor to seminars, conferences, and congresses.

Alexandre began his career at Procter and Gamble, where he gained executive experience in manufacturing, quality assurance, and logistics in assignment in Brazil and in Europe, but for the past 15 years, he has developed his career as consultant. Alexandre has worked as a senior consultant, trainer, and expert advisor for companies such as GE, 3M, Sony, Bayer, Pepsico, Pernod Ricard, Unilever, Avon, Adidas, John Deere, Eaton, Walmart, Motorola, and several others.

Alexandre holds a Bachelor of Arts degree in Chemical Engineering and a Master in Finance degree from the State University of Campinas (Unicamp, Brazil) and a Master of Science (Honors) degree in Logistics and Supply Chain Management from Cranfield's University School of Management, United Kingdom.

Since 2001, Alexandre has actively influenced logistics thinking in several supply chain organizations. He chaired the Logistics Committee at the American Chamber of Commerce (Amcham, Brazil, 2001–2004), chaired the Supply Chain Committee at the British Chamber of Commerce (Britcham, Brazil, 2006–2007), and he was Regional Vice-President of former Brazilian Logistics Association (2002–2003). He has lectured in MBA courses since 2004 in top regional universities such as State University of Campinas (Unicamp) and State University of São Paulo (USP).

Alexandre has published in the *International Journal of Physical Distribution and Logistics Management* and coordinates the Operations and Supply Chain Academic group at LinkedIn, which currently has nearly 20,000 members (March 2014).

Anne Gimeno is a founding-partner at CEBRALOG, a supply chain consultancy and training company headquartered in Brazil since 2001. Anne has also been the Managing-Director of the Brazilian Institute of Supply Chain Professionals (I.B.S.) since 2007.

Anne began her career at Procter and Gamble, where she gained executive experience in materials planning, sales, and customer service. For the past 15 years, she has worked as a senior consultant for companies such as GE, 3M, Bayer, Pepsico, Pernod Ricard, and several others.

Anne holds a Bachelor of Arts degree in Chemical Engineering and a Master of Science (Honors) degree in Logistics and Supply Chain Management from Cranfield University's School of Management, United Kingdom. She served as the director of the former Brazilian Logistics Association (2002–2003).

About I.B.S.

The Brazilian Institute of Supply Chain Management Professionals, a leading regional professional association since 2007, benefits its members via technical events, courses, and an annual congress in São Paulo, Brazil.

I.B.S. promotes knowledge transfer with other knowledge centers around the world. The Committee for International Cooperation (CCI) is the structure that builds partnerships for technical cooperation and knowledge exchange with foreign institutions. Ideal partners are national or regional professional associations or universities.

You can find more information about I.B.S. at www.ibpsc.net/IBS.

Preface

"Vision without action is a daydream. Action without a vision is a nightmare."

—Unknown

All companies have to deliver results to customers, to shareholders, and to stakeholders. Each one within the organization must be aligned to contribute to this endless effort.

But as supply chains increase in complexity, it has become less clear what results each knowledge area within the organization is expected to deliver. Even worse, it has become unclear whether all members of the organization are really capable of contributing to fulfill either customer or shareholder expectations.

The ability to understand the dynamic nature of modern supply networks and to assign practical actions to selected teams within the business has been lost over the years.

This book presents simple, though efficient, tools and methodologies to guide supply chain professionals in their continuous challenge of delivering better results each day. Enriched with many examples, this work is a reference to map processes, define controls, manage operations, and promote change.

Authors' Note

Over the years, the common understanding of world-class operations has evolved from the simplistic, focused management of functional silos to a comprehensive approach of supply network management as the driver to deliver ultimate shareholder value. Although many commentators have tried to describe this evolution, most have failed to properly address the supply chain's fundamental building block: knowledge management. Therefore, their analyses also overlooked the only element that delivers long-term sustainable shareholder value: people.

We are writing five books for Pearson that cover the most important features of this evolutionary journey. These books will provide detailed roadmaps and models to diagnose, implement, and sustain world-class supply chain network management in organizations of all types:

- *A Guide to Supply Chain Management: The Evolution of SCM Models, Strategies, and Practices* (an e-book) introduces the core concept of knowledge management as the only strategy capable of steering supply chains networks management to successfully compete in highly competitive markets. This introductory work reviews supply chain practice from its earliest stages and presents reference models that support our view of this discipline as a business driver to deliver shareholder value.

  This book introduces the *Supply Network Alignment Reference Model* (SNAR Model), which organizes the supply chain networks into knowledge areas that enable accurate decision making from the strategic level to daily management decisions.

  This book also introduces the *Supply Network Knowledge Management Maturity Roadmap* (SKMap). Before the development of a supply network reference model, it was necessary to understand the intermediate evolutionary stages of knowledge management within the supply chain. The SKMap organizes and correlates several strategies and practices according to a unique structure that allows you to understand how to face the future challenges of managing supply chain networks in fluid and complex environments.

- *Supply Chain Management Strategy: Using SCM to Create Greater Corporate Efficiency and Profits* explores how supply chain management delivers shareholder value. The introduction covers topics such as the supply chain master plan, cash-management cycle, purchase-to-pay cycle, and manufacturing-to-revenue cycle. This book introduces the Supply Network Business Value Model (SNValue Model) and discusses the supply chain mechanisms that generate value for the business. It addresses the following topics: enabling sales volume growth, enabling market-share growth, reducing revenue cycle, reducing lost sales, supporting marketing and sales initiatives, enabling customer experience by improving customer perception, managing the cost to serve, offering differentiated service packages, enabling margin growth, reducing cost of sales, balancing asset management, and balancing service level and cost structure.

  This book also presents the *Business Value Impact Chart* (BV Chart) and the *Balanced Control Panel* (BC Panel). The third part of the book covers how each of the SNAR Model knowledge areas can contribute to each of the factors that enable shareholder value. The tool used to establish these relationships is the BV Chart.

- *Executing the Supply Chain: Modeling Best-in-Class Processes and Performance Indicators* covers the supply network governance cycle and explains the mechanisms needed to understand the business though process mapping, risk analysis, and the definition and use of performance indicators for all areas directly or indirectly related to supply chain management. The second part of the book presents how each of the SNAR Model knowledge areas can be monitored and controlled by performance indicators. Other chapters present real world metrics from companies of different sizes, sectors and countries, and discuss benchmarking techniques.

- *Customer Service Supply Chain Management: Models for Achieving Customer Satisfaction, Supply Chain Performance, and Shareholder Value* focuses on the role of customer service as a strategic integrator for differentiated supply chain management. This book presents the *Customer Service Management Model* (CSM Model), a dynamic mechanism developed to evaluate the interactions present in the customer service environment. The model presents four pillars and provides a quantitative approach to understand the connection between them:

  1. Customer Service Level Expectation
  2. Supplier Service Level: Hired Performance
  3. Customer Service Level Perception
  4. Supplier Service Level: Delivered Performance

  Although the book discusses some traditional customer service elements such as pre-transactional, transactional, and post-transactional service, the most important topics are customer service strategies, managing service levels, and customer service organization, respectively.

- *Managing Supply Chain Networks: Building Competitive Advantage in Fluid and Complex Environments* presents a solid roadmap for managing knowledge within organizations across all industries. You learn how to build, implement, and sustain long-term knowledge management as a consistent strategy to deliver business value through supply chain innovation leadership.

  This book presents the *Supply Network Governance Diamond Model* (SNG Diamond) which is executed through…people!

The SNG Diamond Model is a common governance structure focused on the long-term success of the entire supply network that connects knowledge management and risk management and reviews policies that promote the innovative environment required to face the challenges of managing fluid and complex supply networks

1

Supply Network Governance Cycle

Despite the need to coordinate efforts within various elements of the supply chain has recently increased, companies do not yet understand the process of building efficient performance indicators for their operations knowledge areas.

Companies pursue continuous improvement and monitor performance through a set of lengthy and diverse indicators that allegedly represent a process or a cluster of processes.

Performance indicators should be used as decision supporting tools. For that reason, some attributes are expected:[1]

- They have to be quantitative. No performance output should be identified by words such as *good, regular, bad, weak, partial,* or *equivalent.* Metrics must be numbers—either absolute or relative (percentages).
- Performance indicators should promote the right behavior and appropriate organizational culture; therefore, it is desirable for a multifunctional approach to define them.
- Performance indicators must balance the effort to collect and analyze data and the benefits this indicator delivers to the business.

[1] Source: Adapted from "Keeping Score: Measuring the Business Value of Logistics in the Supply Chain," CSC, University of Tennessee.

- Performance indicators must be visible to people, accepted by the organization, and shared within various departments.
- Only what is really important should be measured. The challenge is to define a short list of metrics that are relevant and capable of representing the health of the processes.
- Performance indicators are sensitive to process performance oscillation. A good performance indicator reacts according to minimal process performance deviation.
- The performance indicator must be easily understood by all stakeholders.

Ideally, any performance indicators have all attributes described here. This book introduces a methodology to support companies to define a precise and valuable set of metrics entitled the Supply Network Governance Cycle (SNG Cycle).

The SNG Cycle proposes a five-step roadmap that guides companies from the early stages of acquiring fundamental business visibility through process mapping to advanced people management policies. These five steps are as follows:

1. Define the scope.
2. Map the process.
3. Analyze the risks.
4. Define the metrics.
5. Review the SNG Cycle.

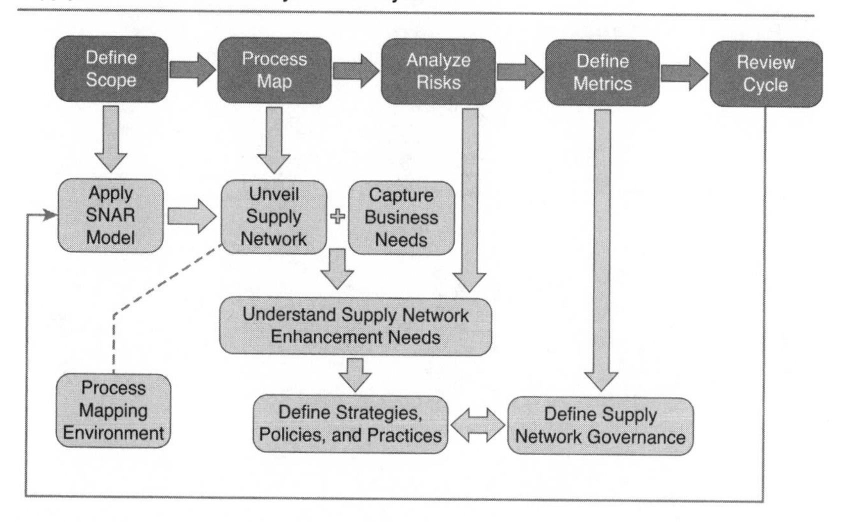

Figure 1.1 Supply Network Governance Cycle

1.1 Scope Definition

The mechanism the Supply Network Governance Cycle uses to identify the exact process scope (set of activities) to be mapped is the Supply Network Alignment Reference Model (SNAR Model) and its coding system (Oliveira and Gimeno 2014_01).

The SNAR Model provides a structured approach to map the extended supply network from different perspectives. These perspectives may vary both in scope (number of functions) and depth (functional details), using as an example a customer service team that is to map and monitor the transportation processes and its impact on several key customers.

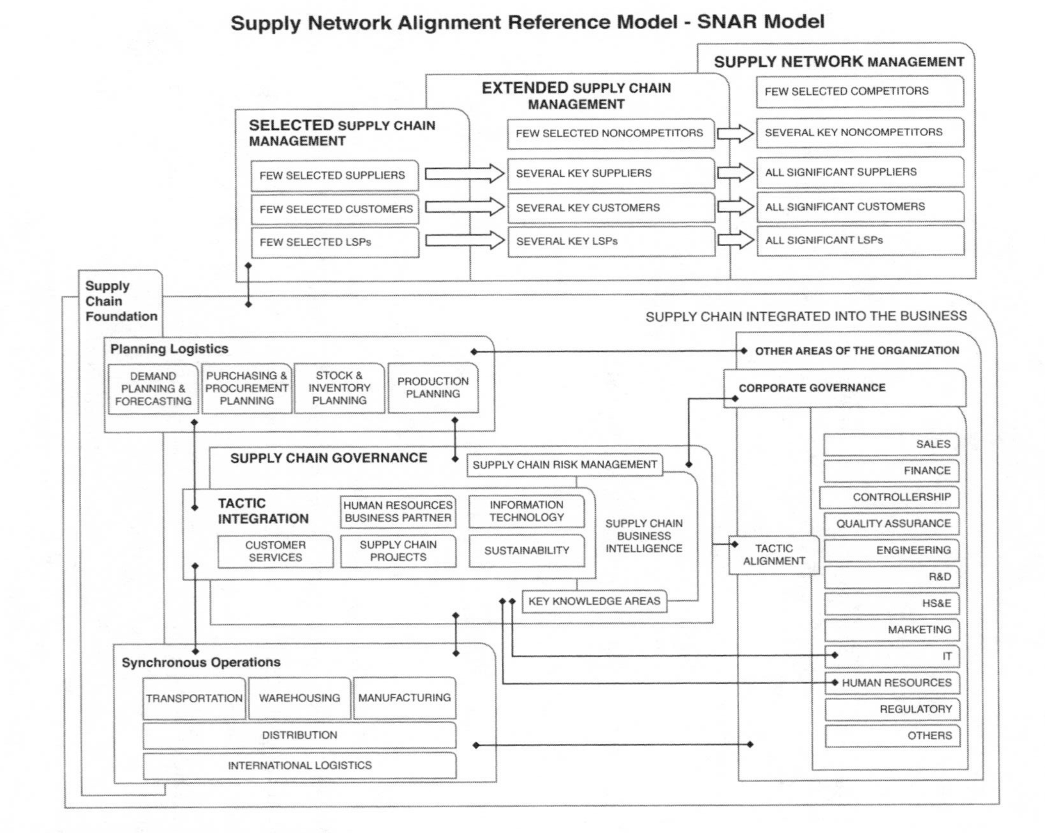

Figure 1.2 SNAR Model

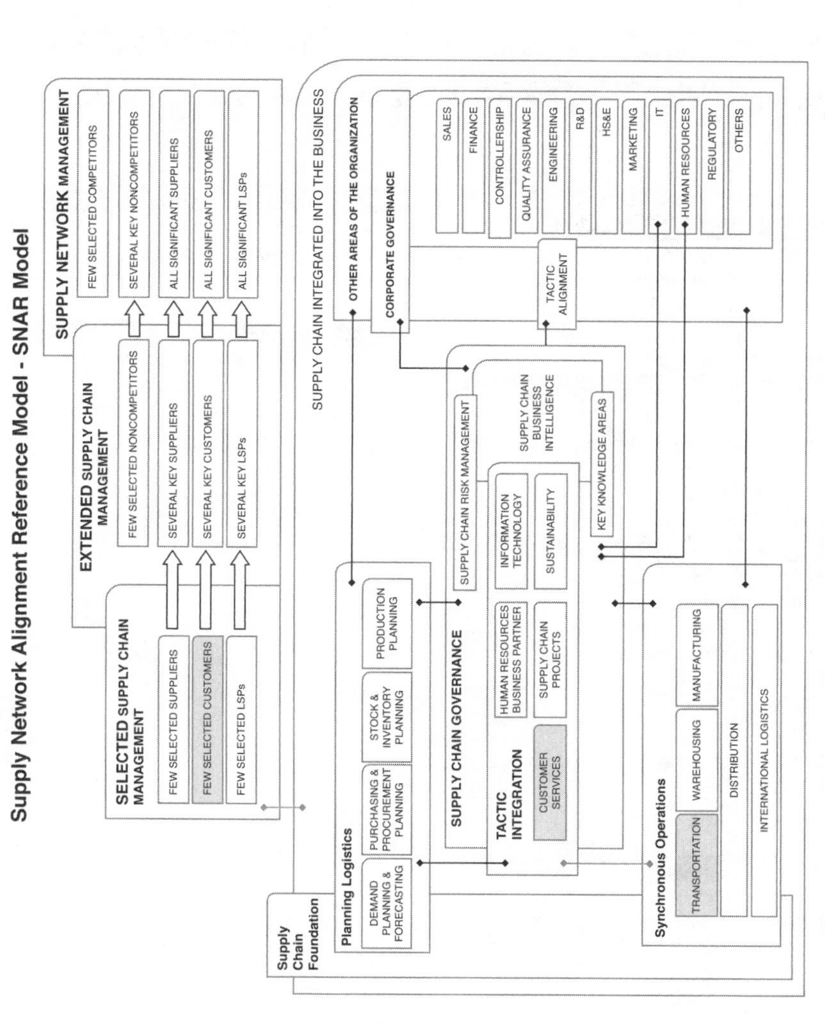

Figure 1.3 SNAR Model Customer Service and Transportation

A more specific approach identifies this environment according to the SNAR Model codification system: 01.02.01 (Transportation), 01.03.01 (Customer Service), and 02.01.02 (Preferred Customers).

01	INTERNAL NETWORK
01.01	Planning Logistics
01.01.01	Demand Planning and Forecasting
01.01.02	Procurement & Purchase
01.01.03	Stock and Inventory Control
01.01.04	Production Planning
01.02	Synchronous Operations
01.02.01	Transportation
01.02.02	Warehousing
01.02.03	Manufacturing
01.02.04	Distribution
01.02.05	International Logistics
01.03	Tactic Integration
01.03.01	Customer Services
01.03.02	Supply Chain Projects
01.03.03	Information Technology
01.03.04	Human Resources
01.03.05	Sustainability
01.04	Other Departments
01.04.01	Sales
01.04.02	Finance
01.04.03	Controllership
01.04.04	Quality Assurance
01.04.05	Engineering
01.04.06	R&D
01.04.07	HS&E
01.04.08	Marketing
01.04.09	IT
01.04.10	Human Resources
01.04.11	Regulatory
01.05	Supply Chain Governance
01.05.01	Key Knowledge Areas
01.05.02	Supply Chain Business Intelligence
01.05.03	Supply Chain Risk Management

02	EXTERNAL NETWORK
02.01	Preferred Supply Chain
02.01.01	Preferred Suppliers
02.01.02	Preferred Customers
02.01.03	Preferred Service Providers
02.02	Extended Supply Chain
02.02.01	Selected Suppliers
01.02.02	Selected Customers
01.02.03	Selected Service Providers
01.02.04	Preferred Noncompetitors
02.03	Supply Network Management
02.03.01	All Significant Suppliers
02.03.02	All Significant Customers
02.03.03	All Significant Service Providers
02.03.04	Selected Noncompetitors
02.03.05	Preferred Competitors

Figure 1.4 SNAR Model coding system

The next example applies the methodology to the Demand Planning and Forecasting Knowledge area (SNAR 01.01.01).

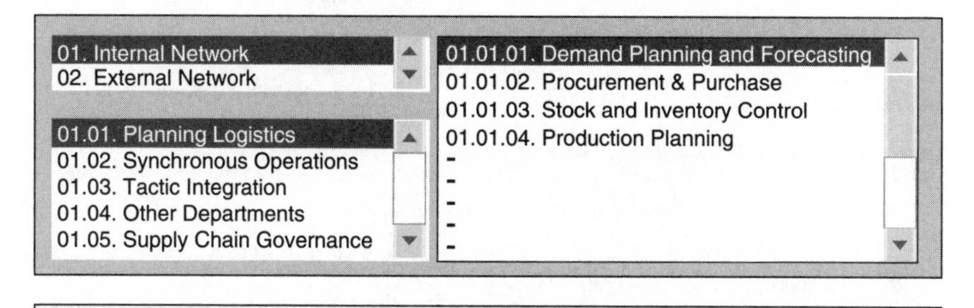

Figure 1.5 Selecting knowledge area

Once the knowledge area is selected, it is required that the process to be mapped has its scope detailed. The following example indicates two key subprocesses:

1. Finished goods demand forecasting [SNAR 01.01.01.a]
2. Sales & operations planning [SNAR 01.01.01.b]

The control chart is a tool that identifies the evolution in each step of the SNG Cycle. The rightmost column indicates the score each step can achieve; the maximum overall score is 16 points.

Code	01.01.01	Processes	Demand Planning and Forecasting	02-feb

Key Subprocesses (KSP):		Process Owners:
KSP a	01.01.01.a Finished goods demand forecasting	Dem.Planning Manager: T.Miros
KSP b	01.01.01.b Sales & operations planning	Team Members:
KSP c		S.J.Thomas, A.Xavier, T.Yian
KSP d		
KSP e		

Supply Network Governance Cycle

Not started = 0.00 Started = 0.25 Completed = 1.00

	KSP a	KSP b	KSP c	KSP d	KSP e	Scale
1. Define scope (apply SNAR model)	1.00					1.0
2. Process Map						
2.a Interviews						1.0
2.b Process observation						1.0
2.c Data analysis						1.0
2.d Organize data						1.0
2.e Elaborate flows						1.0
3. Understand business needs						1.0
4. Define Supply network improvement needs						1.0
5. Elaborate strategies						1.0
5.a Define metrics for strategies						1.0
6. Elaborate Policies						1.0
6.a Define metrics for policies						1.0
7. Define practices						1.0
7.a Define metrics for practices						1.0
8. Consolidate governance strategy						1.0
9. Implement cycle revision methodology						1.0
TOTAL SCORING	1.00	-	-	-	-	16.0

KSP a		*6%*
KSP b		
KSP c		
KSP d		
KSP e		
Scale		*100%*

Figure 1.6 SNG Cycle, control chart

1.2 Mapping the Processes

The step of "mapping the process" is usually misunderstood to be the activity of drawing the process flows. Although the flows play an important role during any process mapping approach, alone they do not offer visibility to the supply network mechanisms.

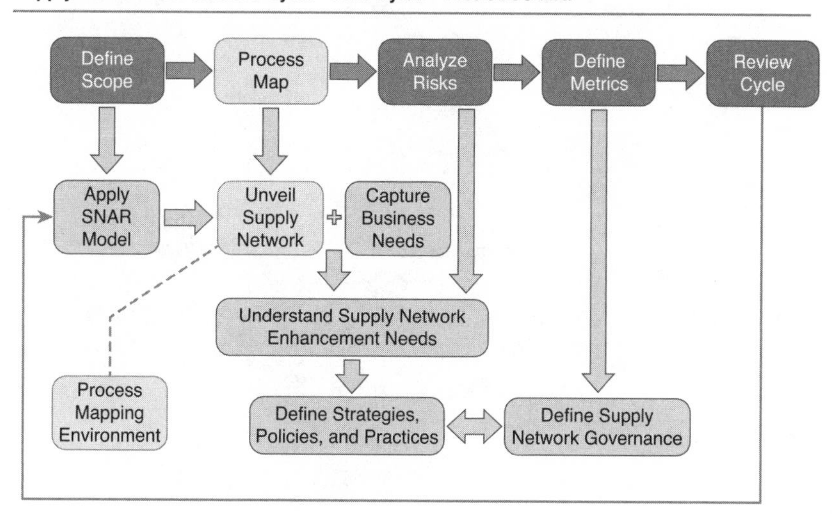

Figure 1.7 SNG Cycle, process map

Managing the process mapping environment requires three core primary and coordinated activities: interviewing people, observing the process, and analyzing data. The next sections explain the elements illustrated in Figure 1.8.

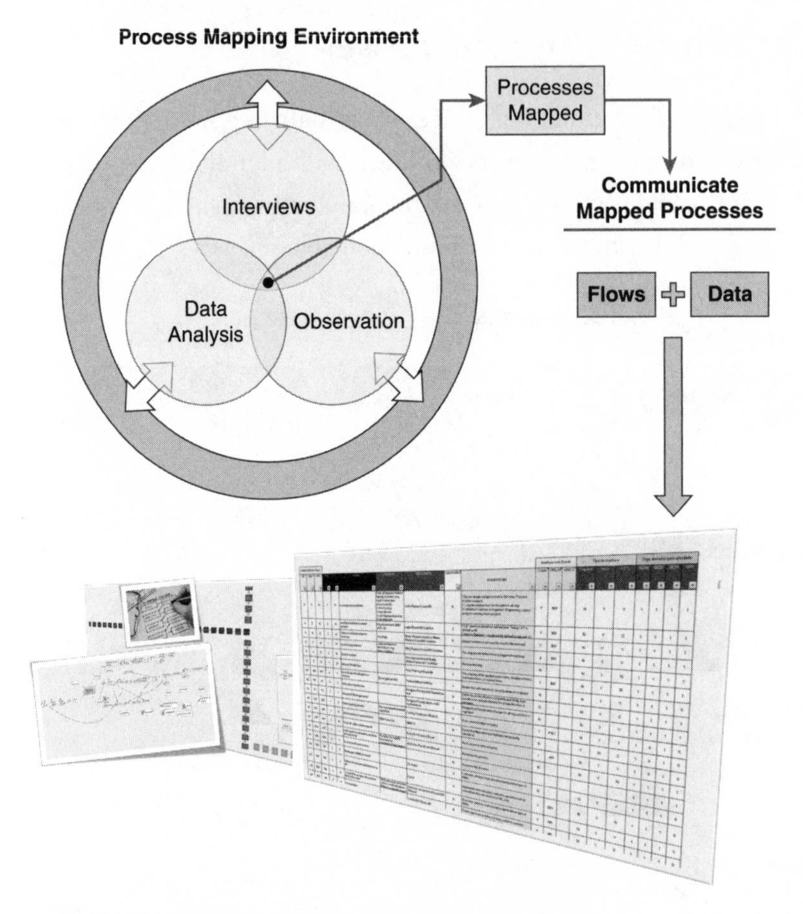

Figure 1.8 SNG Cycle, process mapping environment

These three core activities should follow some basic guidelines built after repeating experiences in different organizations. These guidelines are as follows:

- **Interviewing principles:** (1) One interviewee at a time; (2) open questions (avoid yes-no approach); (3) ongoing record and validation; (4) request complementary data; (5) draw preliminary flow during interviewing process; and (6) double-check all information received.

- **Observation principles:** (1) Information captured during initial interviews should be double-checked before observation starts; (2) observe the process in different moments; (3) open questions to process owners and team; (4) request complementary data; (5) use observations to revalidate preliminary flows drawn during interviews; and (6) double-check observation inputs whenever required.
- **Data analysis principles:** (1) Plan before requesting data; (2) understand the data received and evaluate whether it is exactly what had been required; (3) understand the context and check data adherence to the process; (4) analyze data; (5) double-check the analysis; and (6) revalidate flows previously drawn based on analysis outputs.

Following these three activities the organization needs to organize and communicate the acquired knowledge. A communication strategy is created and it is usually supported by two mechanisms: flowcharts and a supporting database (usually in the format of spreadsheets).

Once the process is mapped, the governance leaders need to capture real information on current business priorities (Figure 1.7, building block 5). Only by cross-checking the process maps and the business key demands is it possible to understand the supply network's enhancement requirements.

The visibility of the gaps between the current process standards ("as is" scenario) and the expected performance ("to be" scenario) is the basis for enabling an appropriate risk analysis.

The evolution within different subprocesses can be simultaneously registered in the control chart introduced in Section 1.1. The following example explores this feature using the Warehousing knowledge area (SNAR 01.02.02).

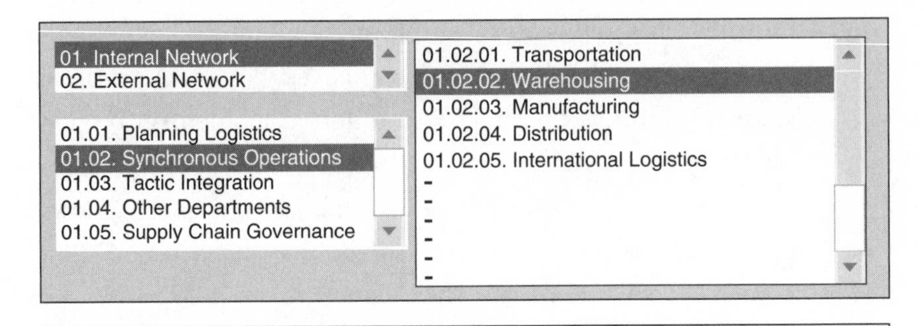

01. Internal Network > 01.02. Synchronous Operation > 01.02.02. Warehousing

Figure 1.9 Selecting SNAR 01.02.02

As explained, it is now necessary to detail which key subprocesses (KSPs) are to be observed during the SNG Cycle. The following subprocesses have been selected: (01) Receiving operations, (02) Picking operations, and (03) Shipping operations. Note that in Figure 1.10 that all three key subprocesses have an equal score (6.0).

In contrast to the previous example, the mapping process and the definition of controls may be at different stages for different subprocesses.

In this example, the organization focuses on the extended supply network and the connections with all significant suppliers (SNAR 02.03.01), customers (SNAR 02.03.02), and service providers (SNAR 02.03.03). The management of the supply network also looks at selected noncompetitors (SNAR 02.03.04) and some preferred competitors (SNAR 02.03.05).

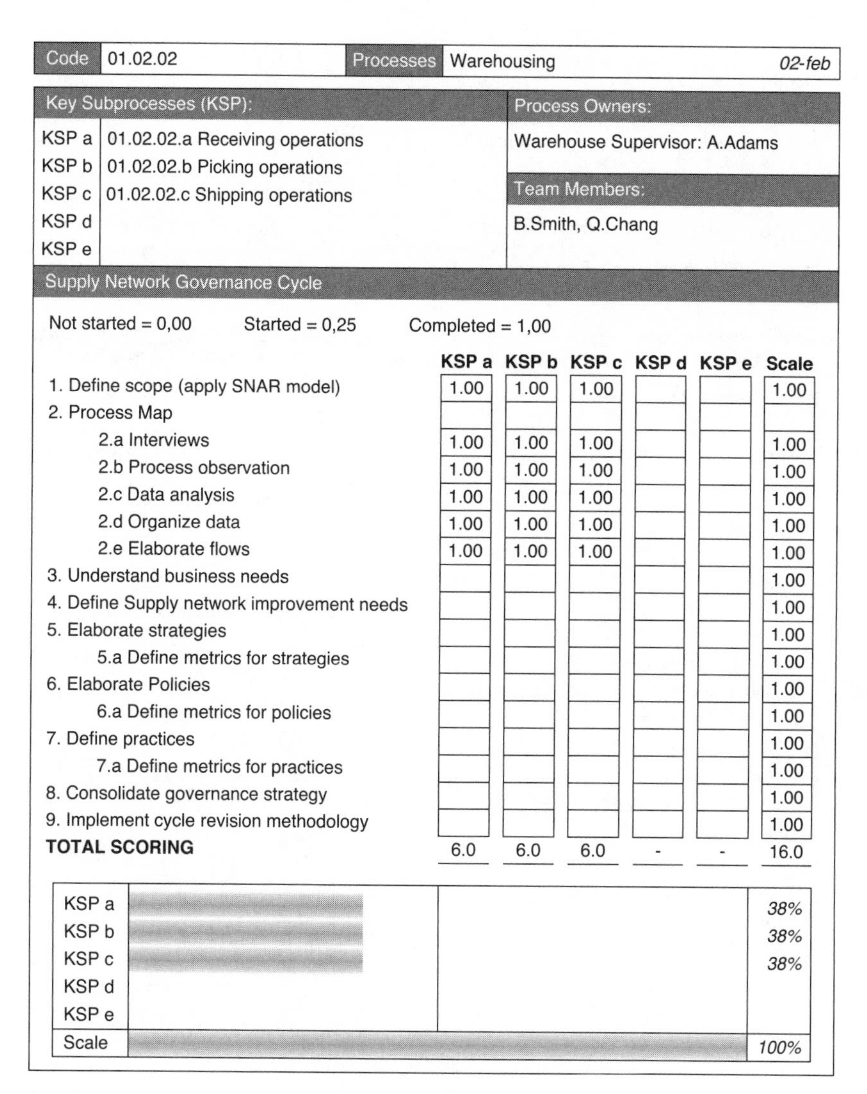

Code	01.02.02	Processes	Warehousing	02-feb

Key Subprocesses (KSP):		Process Owners:
KSP a	01.02.02.a Receiving operations	Warehouse Supervisor: A.Adams
KSP b	01.02.02.b Picking operations	
KSP c	01.02.02.c Shipping operations	Team Members:
KSP d		B.Smith, Q.Chang
KSP e		

Supply Network Governance Cycle

Not started = 0,00 Started = 0,25 Completed = 1,00

	KSP a	KSP b	KSP c	KSP d	KSP e	Scale
1. Define scope (apply SNAR model)	1.00	1.00	1.00			1.00
2. Process Map						
2.a Interviews	1.00	1.00	1.00			1.00
2.b Process observation	1.00	1.00	1.00			1.00
2.c Data analysis	1.00	1.00	1.00			1.00
2.d Organize data	1.00	1.00	1.00			1.00
2.e Elaborate flows	1.00	1.00	1.00			1.00
3. Understand business needs						1.00
4. Define Supply network improvement needs						1.00
5. Elaborate strategies						1.00
5.a Define metrics for strategies						1.00
6. Elaborate Policies						1.00
6.a Define metrics for policies						1.00
7. Define practices						1.00
7.a Define metrics for practices						1.00
8. Consolidate governance strategy						1.00
9. Implement cycle revision methodology						1.00
TOTAL SCORING	6.0	6.0	6.0	-	-	16.0

KSP a	38%
KSP b	38%
KSP c	38%
KSP d	
KSP e	
Scale	100%

Figure 1.10 Control chart, subprocesses

Code	02.03.	Processes	Supply Network management	02-feb

Key subprocesses (KSPs):		Process owners:
KSP a	All significant suppliers	SC Director: A. Gimes
KSP b	All significant customers	
KSP c	All significant service providers	Team members:
KSP d	Selected noncompetitors	C. Glows, P. Zizau, N. Prix
KSP e	Preferred competitors	

Figure 1.11 Selecting knowledge area, SNAR 02.03

The following figure shows the first key subprocess (KSP a: all significant suppliers) has been through all the SNG Cycle, whereas KSP e (preferred competitors) is in its early stages. This company is more comfortable in addressing its suppliers or its customers (KSP b), but the evaluation of significant service providers (KSP c) still requires the definition of metrics to monitor adherence to selected strategies.

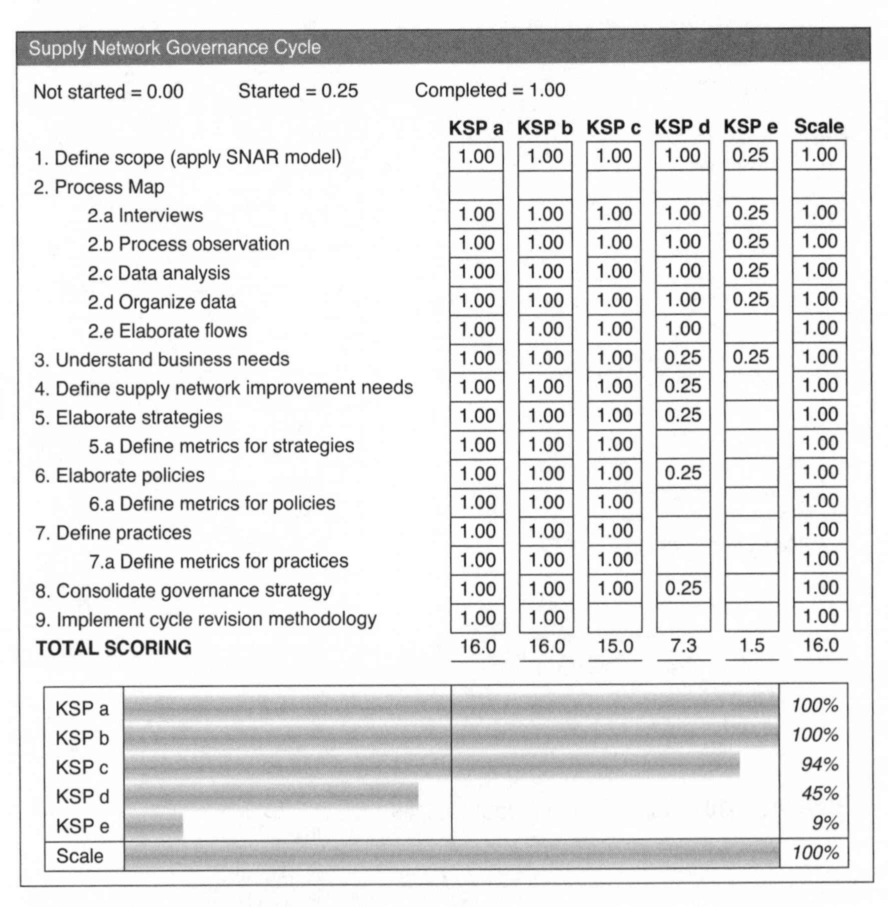

Supply Network Governance Cycle

Not started = 0.00 Started = 0.25 Completed = 1.00

	KSP a	KSP b	KSP c	KSP d	KSP e	Scale
1. Define scope (apply SNAR model)	1.00	1.00	1.00	1.00	0.25	1.00
2. Process Map						
2.a Interviews	1.00	1.00	1.00	1.00	0.25	1.00
2.b Process observation	1.00	1.00	1.00	1.00	0.25	1.00
2.c Data analysis	1.00	1.00	1.00	1.00	0.25	1.00
2.d Organize data	1.00	1.00	1.00	1.00	0.25	1.00
2.e Elaborate flows	1.00	1.00	1.00	1.00		1.00
3. Understand business needs	1.00	1.00	1.00	0.25	0.25	1.00
4. Define supply network improvement needs	1.00	1.00	1.00	0.25		1.00
5. Elaborate strategies	1.00	1.00	1.00	0.25		1.00
5.a Define metrics for strategies	1.00	1.00	1.00			1.00
6. Elaborate policies	1.00	1.00	1.00	0.25		1.00
6.a Define metrics for policies	1.00	1.00	1.00			1.00
7. Define practices	1.00	1.00	1.00			1.00
7.a Define metrics for practices	1.00	1.00	1.00			1.00
8. Consolidate governance strategy	1.00	1.00	1.00	0.25		1.00
9. Implement cycle revision methodology	1.00	1.00				1.00
TOTAL SCORING	16.0	16.0	15.0	7.3	1.5	16.0

Figure 1.12 Control chart, scoring

The elaboration of flows occurs as interviews, observation, and data analysis are conducted. This is a nonlinear process, and interactions of several types may be required.

This methodology also requires that diverse people are part of the process. It is important to take into account the hierarchical and functional barriers that create disconnected islands of knowledge.

From the beginning of this process until the validated flows are finished, the organization need to recurrently go through interviewing, data analysis, or *in loco* observation.

It is possible to structure these dynamics according to six basic stages:

1. Start to map
2. First round flows
3. Preliminary flows
4. Refined flows
5. Validated flows
6. Organized data

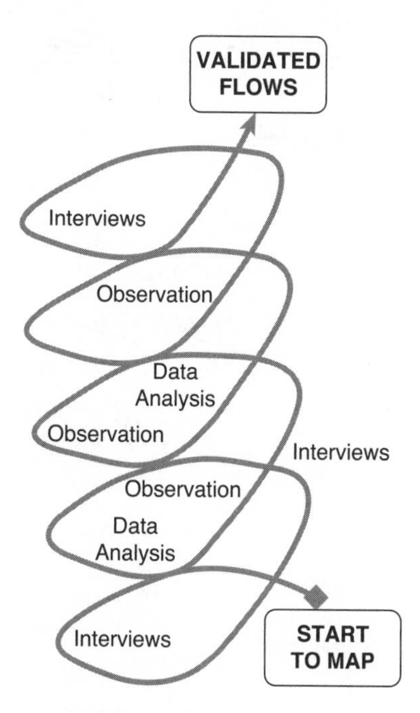

Figure 1.13 Process mapping spiral

The initial three stages are highly dependent on the interviewing sessions, which should consider interviewees with complementary knowledge, whereas information cross-checking on specific processes or activities may be left for a future occasion.

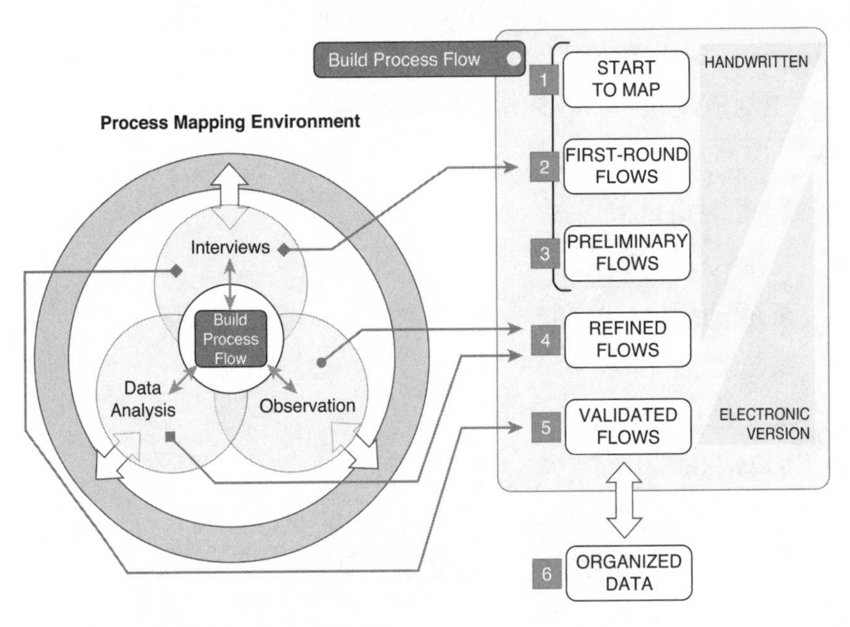

Figure 1.14 Process mapping details

This methodology has been applied to several consulting projects, and the period from initial interviewing to validated flows when supported by properly organized data varied from a few weeks to several months, depending on the scope range and depth. Figure 1.15 summarizes this feature based on some of the projects executed by the authors from 2001 to 2014.

Client Company	Headquarters	Processes Mapped	Mapping Period (months)
3M	USA	3	2.0
AGCO	USA	8	2.5
AJR	BRAZIL	2	2.0
BEIERSDORF	GERMANY	4	2.0
CAMOZZI	ITALY	2	1.0
DANISCO	DENMARK	1	1.5
DRAKA-COMTEQ	HOLLAND	2	1.0
EAGLEBURGMANN	GERMANY	1	1.0
FIRMENICH	SWITZERLAND	3	2.0
FLEXTRONICS	USA	6	2.5
FMC	GERMANY	3	2.0
GE	USA	9	4.0
HUSSMAN	USA	4	2.5
HYPERCOM	USA	7	3.0
KATOEN NATIE	BELGIUM	12	3.5
MAGNA CLOSURES	CANADA	5	3.0
MARCOPOLO	BRAZIL	1	1.0
PERNOD RICARD	FRANCE	1	2.0
SONEPAR	FRANCE	3	1.5
SONY	JAPAN	1	1.7
ZODIAC	ARGENTINA	2	2.0

Figure 1.15 Process mapping period

Although it sounds curious, the best tip to give to someone who needs to draw a flow for any given process is this: Start it! In fact, the only resources you need are pencil, paper, and an interviewee.

Figure 1.16 is the result of a 1-hour conversation with a customer service manager in a beverage company. The session focused on the order management process and identified several key activities, some of which are listed after the image.

This initial session identified a total of 34 elements. As the process mapping occurred, complementary data was analyzed, and observation happened; this initial picture evolved into a final validated version with 135 activities identified.

The methodology used to draw the flows may vary according to the final goals. Some of the best-known business process mapping methodologies[2] are, in alphabetic order, business model canvas, business process discovery, business process modeling, ethnography, IDEF, OBASHI, N2 Chart, organizational studies, process-centered design, SADT, systems engineering, value stream mapping, and workflow.

Once early, unconnected flows generated during first interviews are available, the mapping team can create more comprehensive diagrams. Although some teams would prefer to go electronic, this phase is usually done with pencil and paper because there is still much to find out about the process and rework is likely to be necessary before any fine-tuning activities.

The first round of flows builds a scenario similar to a puzzle. Often, a wide panel (or wall) is used to offer a complete view of the processes being mapped. Recent and old versions are usually side by side to facilitate overall visibility.

[2] http://en.wikipedia.org/wiki/Business_process_mapping, February 2014

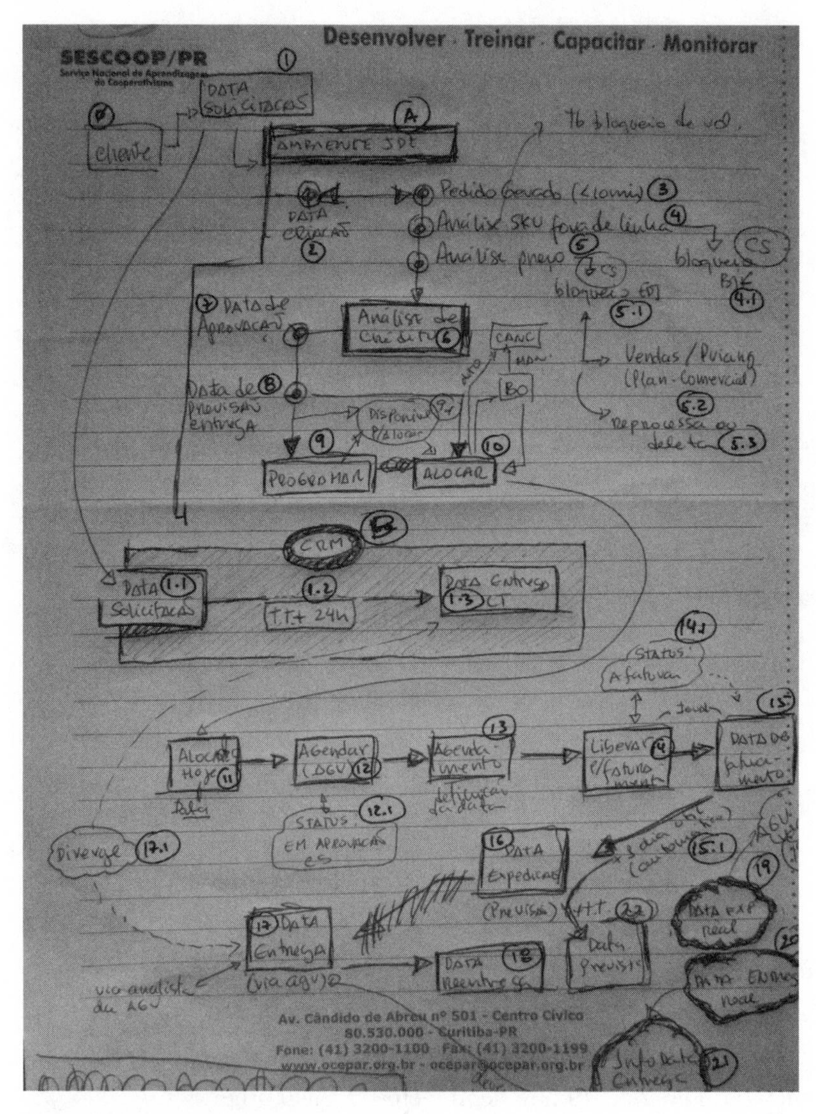

Figure 1.16 Process mapping, start

Some elements:

- 0 Client
- 1 Order request date
- 2 Order creation date
- 4 SKU analysis

- 5 Price analysis
- 6 Credit analysis
- 8 Define delivery date
- 12 Schedule delivery
- 15 Invoicing
- 16 Shipping
- 21 Delivery confirmation

Figure 1.17 First-round flows

Depending on the magnitude of the mapping efforts, a "war room" may be used to create a focused environment. More comprehensive projects involving larger teams and longer schedules will benefit from the structure of a war room. Figure 1.18 shows the war room used for a project at GE Transportation. This team worked simultaneously on 9 maps and created 49 flows connected to each other.

Some complementary interviews are expected during the phase of drawing the first round of flows. Additional sessions are usually required before the preliminary flows are done. The goal of preliminary flows is to organize all information captured so far in a format detailed enough to go into a thorough presentation to those who were interviewed. The next figure shows a war room where processes are organized on different walls. These maps were to go through a two-day session with the interviewees.

Figure 1.18 War room

Figure 1.19 Preliminary flows

Following the adjustments resulting from the presentation of the preliminary flows to the interviewees, it is now time to complete the process diagnosis with *in loco* observation and data analysis. Each project is unique. Therefore, the selection of what data must be analyzed or what processes need observation do not follow a standard pattern. The process mapping team frequently plays detective in this phase.

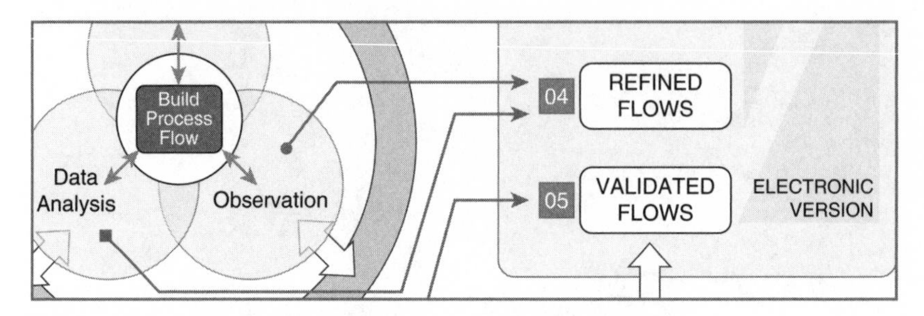

Figure 1.20 Refining flows, data analysis, and observation

Any process mapping must be executed with a clear purpose. The diagnosis will always be more efficient when a target has been defined in the early stages. Examples of these targets include the identification of:

- Low-productivity activities
- Activities with an interface with information systems
- Processes that generate inventory divergence
- Processes that cause customer dissatisfaction
- Any other approach that will improve organizational performance in any area of the business

There are infinite possible targets to a process mapping exercise.

Business Case

This process mapping was chosen to understand the root causes for recurrent inventory divergence in an external manufacturing facility that serves a large multinational electronics industry headquartered in the United States. This case is inspired by an operation that includes three mobile assembly lines, one parts warehouse, and one finished goods distribution center.

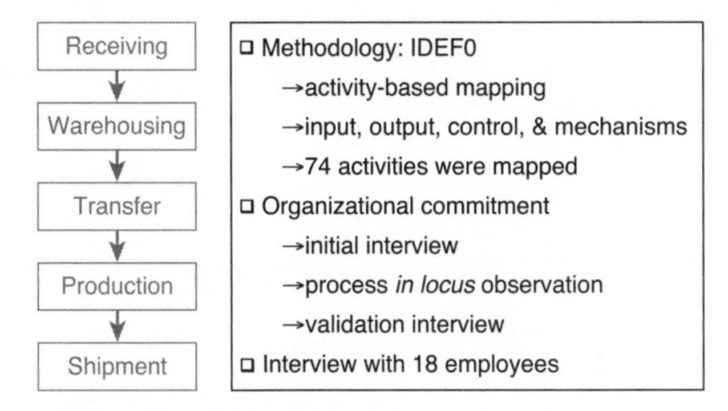

Figure 1.21 Case, scope

The mapping methodology used was IDEF0. Five processes were mapped: receiving, warehousing, transfer, production, and shipment. A total of 18 employees were interviewed and 74 activities identified. The goal of the mapping was to identify the cause of recurrent inventory divergence.

Each of the five processes was diagnosed after interviews, *in loco* observation, and data analysis. This diagram indicates 5 activities out of 12 are vulnerable and inventory divergence may occur. Note these items are parts for mobile devices. They are very small pieces; some of them can hardly be seen.

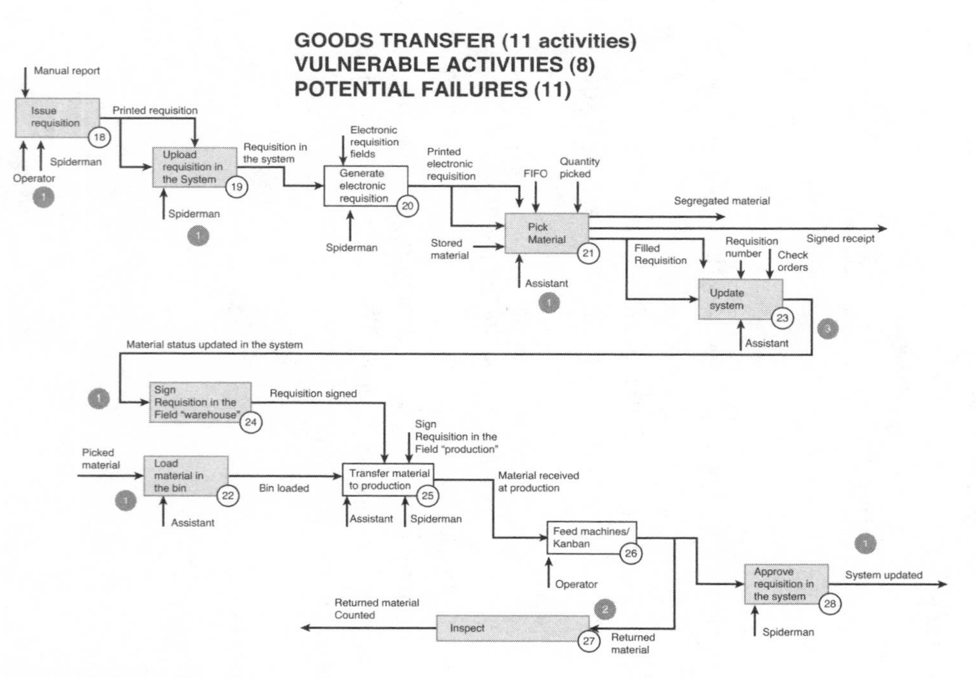

Figure 1.22 Case, map

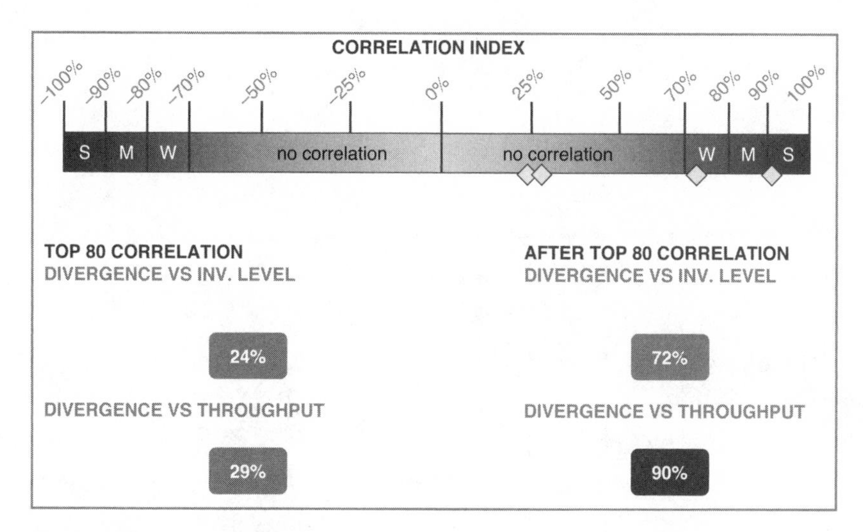

Figure 1.23 Case, analysis

A detailed data analysis investigated the eventual correlation between inventory divergence and inventory holding or volume throughput. Throughput was shown to be correlated to the divergence of 80% of the items. These items represented only 45% of the divergence in value.

The challenge is then to identify the reason why 20% of the items generate 55% of the divergence.

Other analyses were done, such as inventory aging and divergence value by shift and by operator.

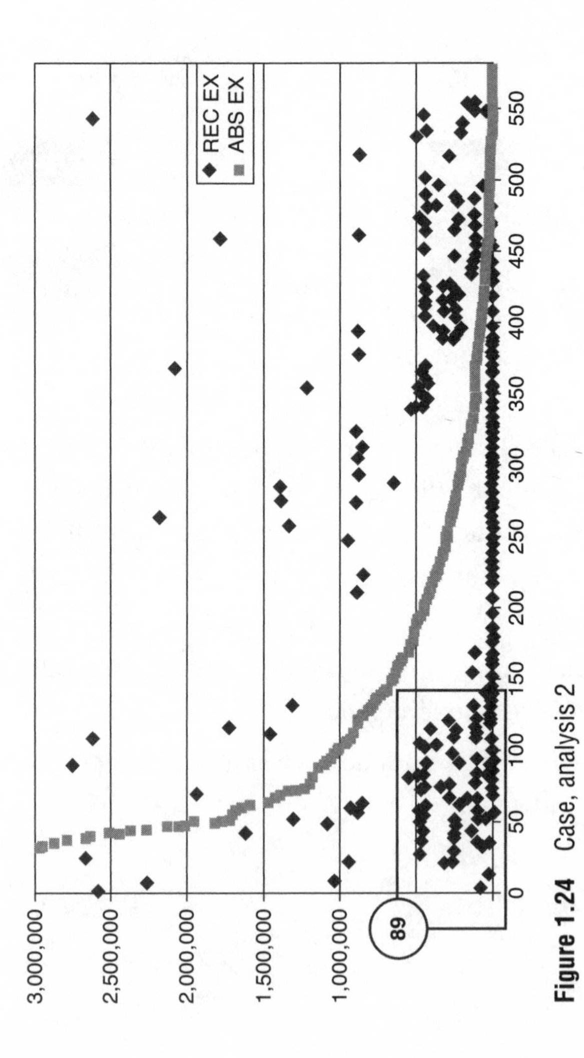

Figure 1.24 Case, analysis 2

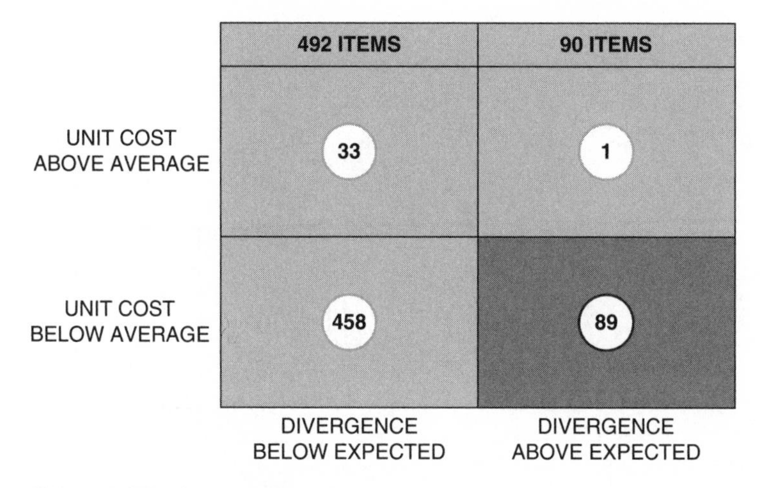

Figure 1.25 Case, analysis 3

A possible cause was the unit value of these items. If these 89 items (20% of total) had a unit cost above the average unit cost, this could explain why their divergence summed such a high value. But all these items with high value divergence had a unit cost below the average unit cost.

It was during the observation *in loco* of the manufacturing operation that one of the consultants noticed a junior operator was sweeping some items that were on the floor of the production area. When the consultant asked where these little items come from, the operator indicated three equipments (as circled in the map).

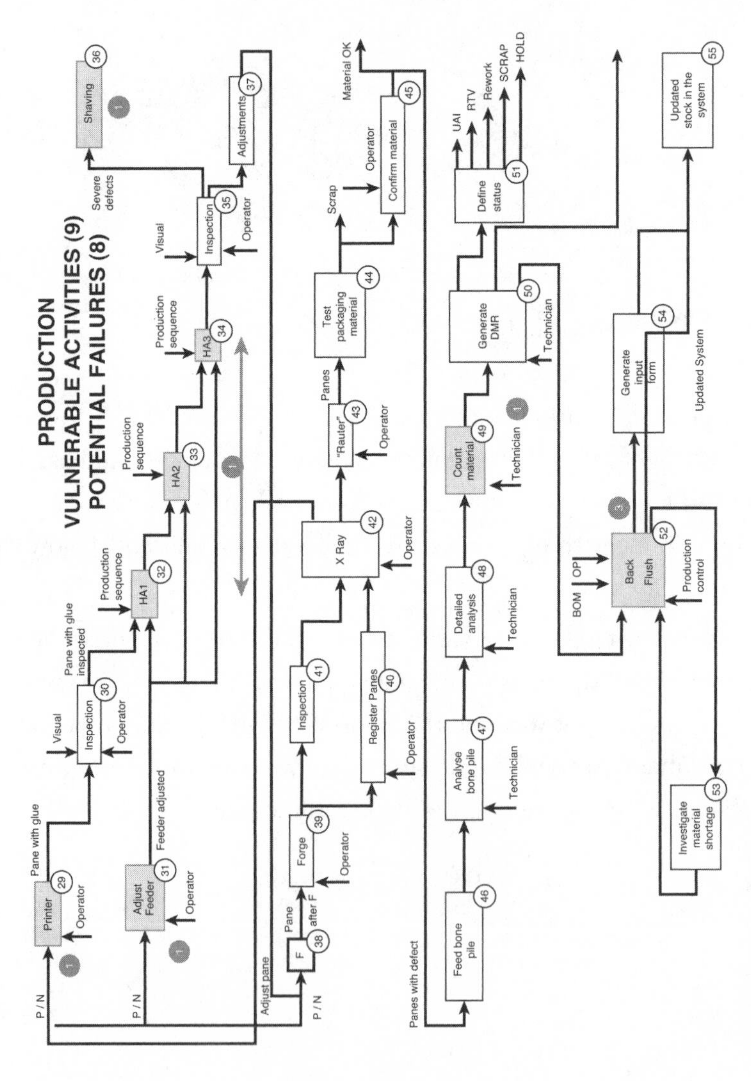

Figure 1.26 Case, diagnosis

Conclusion

These three process equipments identified in the production flow faced some performance issues and several small parts that were processed through them were systematically being scrapped without being properly registered. The company would only acknowledge the missing components during inventory audit, but would not be able to track them back to the root cause. The combination of several factors affected 80% of the items that responded to 45% of inventory divergence. This required process improvements in all mapped areas. And a single factor related to the manufacturing operation caused 55% of divergence by not registering losses on 89 items, which represented 20% of total stock keeping units (SKUs). As a consequence of the process mapping, the following report was issued. Some information has been omitted due to confidentiality reasons.

Project Report: Sample

A) SUMMARY

- There is a weak correlation between the generated divergence and the product quantity in stock.
- A strong correlation exists between the generated divergence and the product quantity put into motion by the company.
- The items that generated more divergence (top 80) don't fit into the patterns of correlation cited above, so they require better research.
- Some items have a higher % of losses than expected.
- Items with a % of losses higher than the expected can be originated by production activities.
- There is strong evidence of failures in the records of material transferences for the production and from the process of finished goods audit (FGA).

- There are generalized failures on all stages of material flow, with weak or nonexistent controls.
- There are three systems for the control/identification of divergences:
    1. Cycle count, which is inefficient
    2. FGA, which is a reactive system and therefore inefficient on the prevention of divergences
    3. Physical inventory, an equally reactive system
- There are points of vulnerability in the production associated with not reporting all losses.
- There are failures in the internal communication of operational problems.
- There are conflicts of interest in the hierarchy of decision making, from the top down and the bottom up.
- The impact of failures in the material flows is reflected in the materials divergence on the work in progress (WIP), but is not generated totally in the production

B) Transformation Plan: Basic Steps

- Phase 1: Elaboration/review of approximately 40 procedures (receiving, warehousing, and transference to production). Implementation of actions plans.
- Phase 2: Parallel to phase 1. Elaboration and implementation of continuous training program.
- Phase 3: After phases 1 and 2. Complete review and implementation of an independent cycle count program.
- Phase 4: Parallel to phases 1, 2, and 3. Elaboration or review of approximately 27 procedures (production & expedition). Implementation of action plans for those areas.

- Phase 5. After phases 3 and 4. Implementation of a process-based audit with total independence, on a daily basis, reporting to the site top management.

C) Potential Barriers to Implementation

- Reorganization of the internal communication system
- High turnover of employees
- Overloaded leading team (supervisors/managers)
- Leading team (supervisors/managers) with little experience on:
  - Procedures elaboration techniques
  - Internal audit techniques
  - Cycle count program (design, start-up, rollout, ongoing)
- Many vulnerable activities that must be adjusted simultaneously

D) Loss-Elimination Scenarios

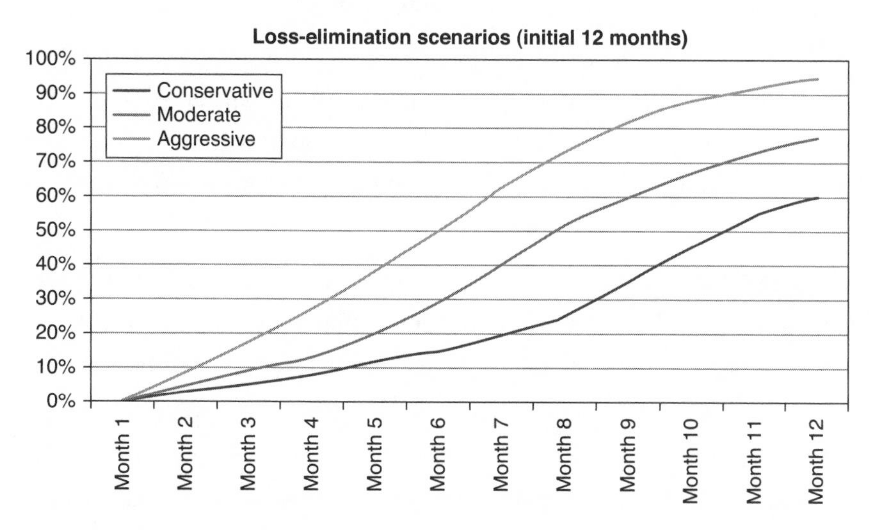

Figure 1.27 Loss-elimination scenarios, 12 months

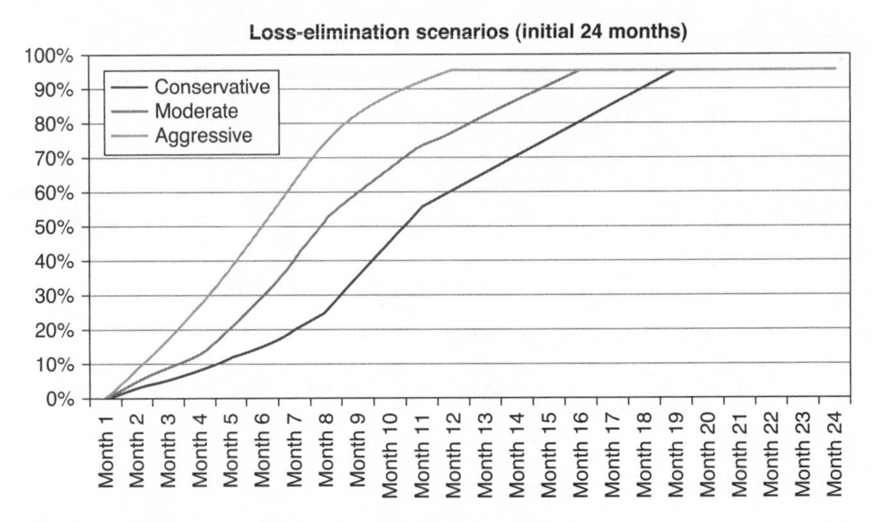

Figure 1.28 Loss-elimination scenarios, 24 months

Validated Flows

At this time, nearly all the information has been captured in various interviews, innumerable data has been analyzed, and all significant processes have been observed *in loco*.

This session differs from previous ones because it is not an interview. Ideally, the interlocutor will only confirm that the presentation is accurate and represents the processes mapped with high fidelity. If relevant, new information comes up at this time; this will indicate that the previous stages have not been properly done.

The format this connected puzzle presents is varied; there is no golden rule. The more information the mapping team has captured, the more resources will be available to illustrate this final flow.

The following image is based on the process mapping of a warehousing operation of a retailer. Note the arrows indicate correlations between different subprocesses. Some graphs (histograms) offer a time-based approach to the process map.

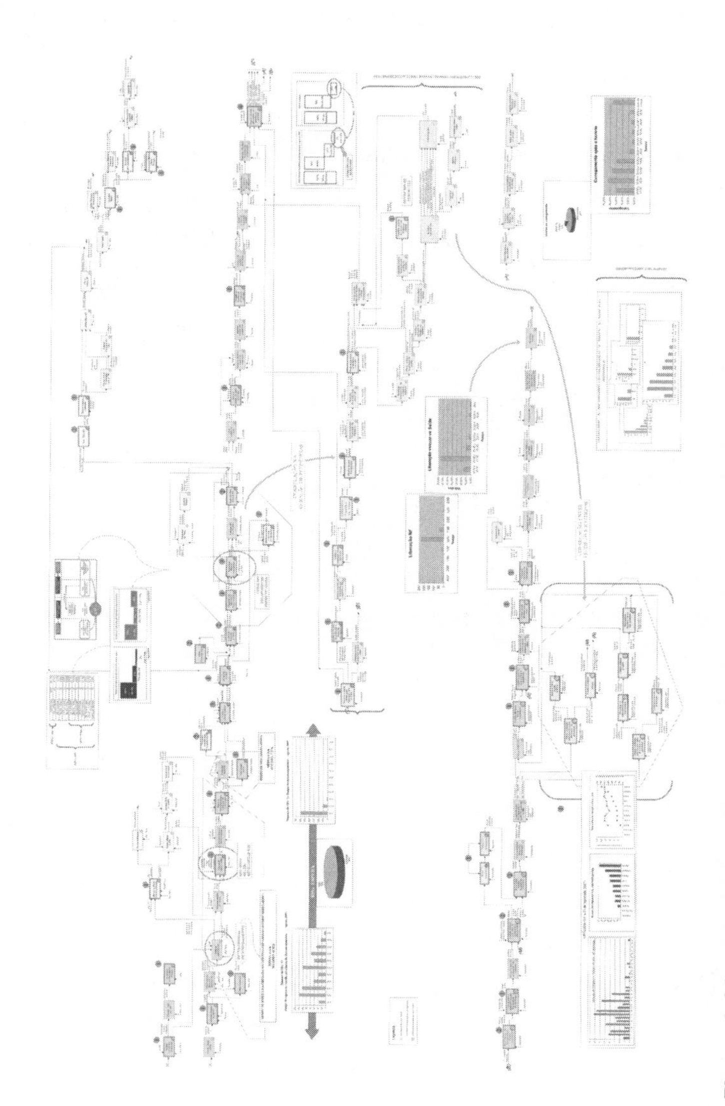

Figure 1.29 Disclaimer: The processes identified in this map will vary based on the organization/industry goals, and so readability (or nonreadability, in this case) is not a factor for example purposes here.

Although the final version of the flow may offer a comprehensive perspective of the mapped processes, it is impossible to converge into it all the information captured during the interviews, data analysis, and *in loco* observations. The use of complementary organized data is therefore fundamental.

Given the popularity of spreadsheets, these are commonly the best way to organize all information captured during the mapping process. The format is highly customized according to the outputs of the project. Note that each line of the spreadsheet represents one specific activity mapped.

# MAP	Area	PRO	n°	Activity	Physical	Intro	CFFT	Interface F/C/F	Critical points	CONTROL	MECHANISM	OBS
A	E	IC	15	Load fiscal attributes & confirm attributes	N	Y	Y	N	Y	Fiscal parameters CEB/CST = do	Layls/ERP IM/Workflow	
L	MP	KITS	5	Explode items of loco BDM & place internal requisition "integrated system"	N	Y	N	Y	Y	For MWS integrated system kit list/99 SKU's	GCM Mtl. Planner/Excel/manual	
AE	CST	DM	5	Start production or repair black flush	Y	N	N	N	Y	MCC/corse aut	MNFCT	
K	MP	GCR	29	Verify item necessity	N	Y	N	Y	Y		Mat. Plan. Org	
K	MP	GCR	30	Transfer item from org to org	N	Y	N	N	Y		Mat. Plan. Org	
M	PUR	PUR DIR	13	Issue PO	N	Y	N	Y	Y	Local item/Direct buy	Buyer	
M	PUR	PUR DIR	24	PO goes through approval levels	N	Y	N	N	Y		Buyer, coordinator, or manager	
W	AP	IS	2	Calculate taxes	N	Y	Y	N	Y		Dascan	
W	AP	IS	9	Inform issues	N	Y	Y	Y	Y		Corretora SISBACEN	
AG	RET	RE	13	Accountability	N	Y	N	N	Y		ERP OM/ERP AR	
AG	RET	RE	14	Reversal of the additional account	N	Y	N	Y	Y		Finance AR manual	Classify
N	IMP	IP	28	Create new req	N	Y	N	Y	Y		Mat purchase	
O	IMP	CC	14	Issue NF-e before material receipt	N	Y	Y	Y	Y	Arrival note in MG not removed to EADI/Import experts	External Consultancy	

Figure 1.30 Example of organized data

1.3 Risk Analysis

By having the processes mapped, a number of vulnerabilities are identified and properly registered, as discussed in the previous section. These vulnerabilities must be classified and prioritized. These dynamics may occur through three different mechanisms:

1. Reactive (focused) approach
2. Proactive (comprehensive) approach
3. Intermediate (business-driven) approach

The reactive approach to process vulnerability understanding (risk analysis) begins with a known issue. For example, a warehousing operation (SNAR 01.02.02) that registers a low picking productivity. This scenario calls for a focused process mapping to understand the activities, controls, and mechanisms mainly associated with picking operations.

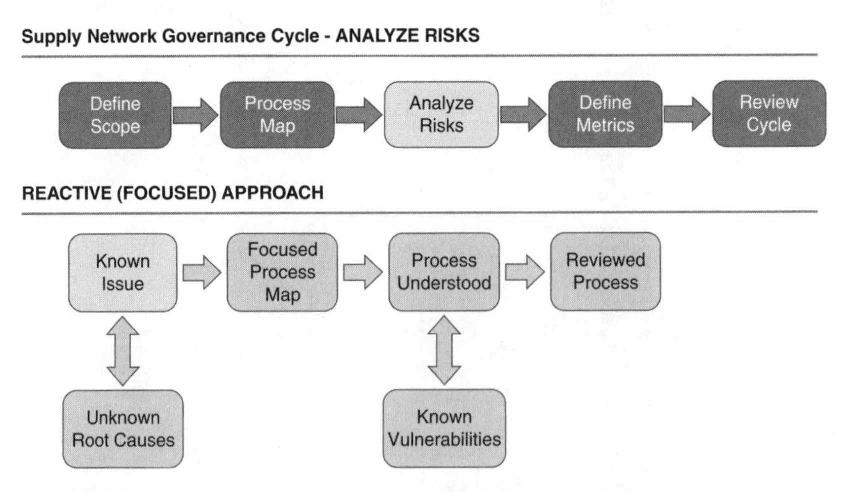

Figure 1.31 Risk analysis, reactive approach

As the specific processes are mapped and the vulnerabilities are understood, process revisions may occur. It is important to remember that process mapping necessarily includes the three pillars:

interviewing people, observing the process, and analyzing data. The difference between actual process and future processes defines the transformation effort.

The proactive approach to risk analysis occurs in two situations:

1. Several diffuse issues have disabled the reactive (focused) approach.
2. It is a planning initiative, prior to the identification of any significant issues, and expects to give visibility to unknown inefficiencies.

The proactive approach begins with a comprehensive process mapping, covering several knowledge areas; it is suggested to follow the SNAR Model for defining the process mapping scope.

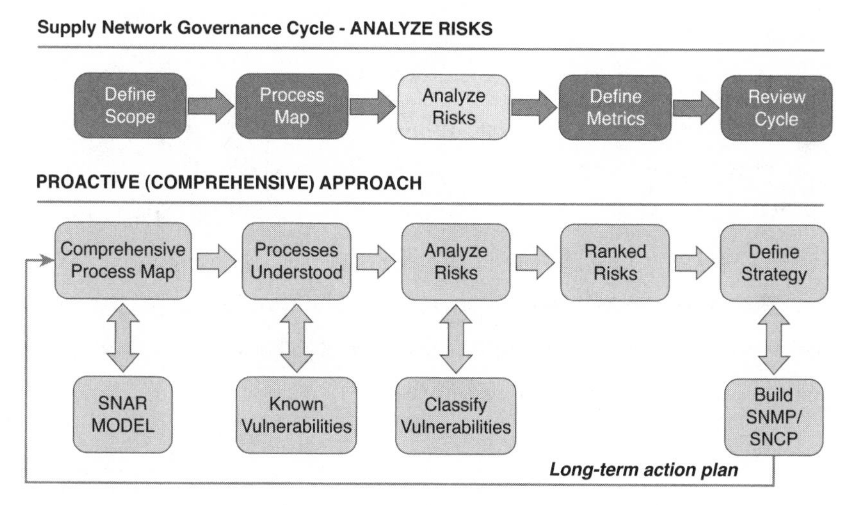

Figure 1.32 Risk analysis proactive approach

After the processes are mapped, innumerable vulnerabilities are exposed that demand prioritization. As the risks associated with vulnerabilities are ranked, it is possible to define the implementation strategy. Depending on the extent of the process mapping, it is possible to build a Supply Network Master Plan (SNMP) as a reference for the transformation process.

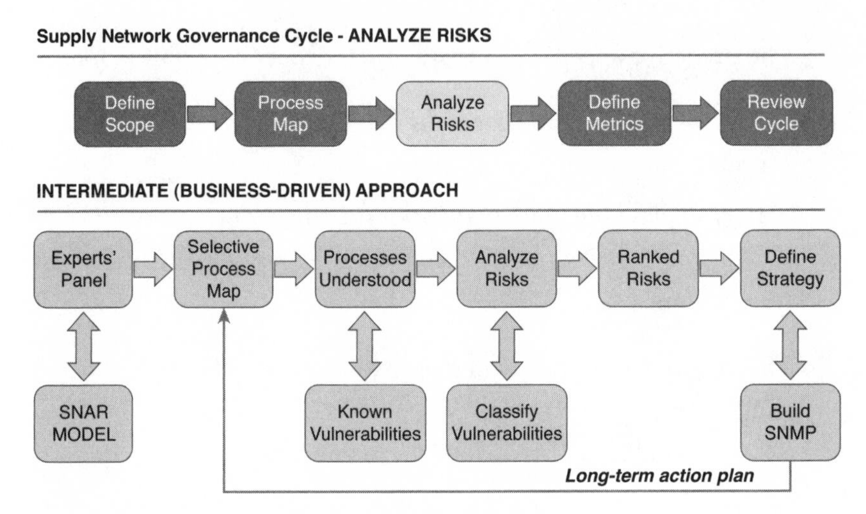

Figure 1.33 Risk analysis, intermediate approach

In addition, the implementation strategy may enable the elaboration of a Supply Network Continuity Plan (SNCP). The intermediate risk analysis approach involves the contribution of experts in the operations to be mapped.

The experts panel selectively reduces the number of processes to be mapped, instead of a comprehensive process mapping a number of selected processes go through the risk analysis.

2

Definition of Metrics

Defining the metrics that govern a process performance is of extreme importance. This is itself a process, which may be illustrated according to a 14-step roadmap:

1. Specify scope.
2. Observe performance.
3. Define performance pattern.
4. Assign a name to the control.
5. Define algorithm.
6. Test algorithm.
7. Define performance goal.
8. Understand learning curve.
9. Map stakeholders.
10. Validate control dynamics.
11. Assign control to people.
12. Implement control.
13. Monitor process-control adherence.
14. Fine-tune control dynamics.

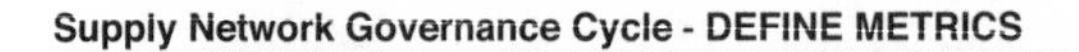

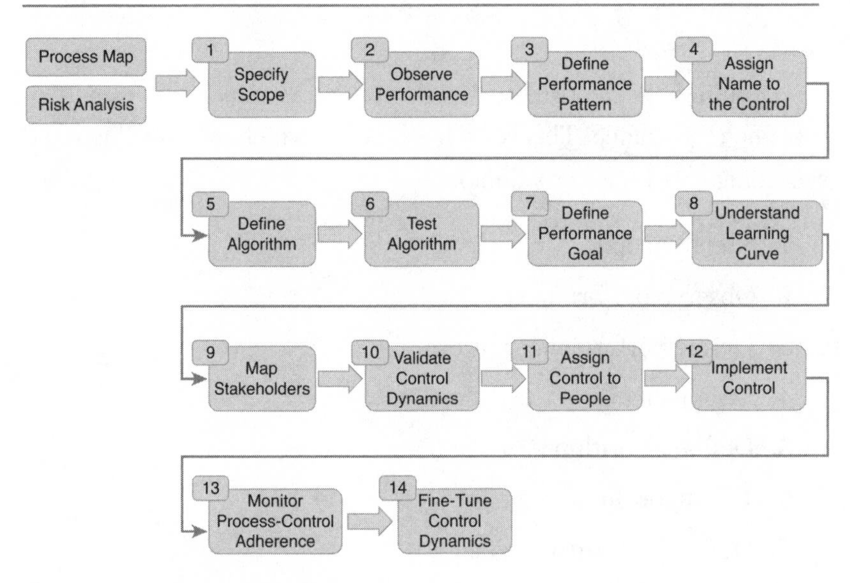

Figure 2.1 Metrics definition

2.1 Specify Scope

Note that this step differs from the scope definition activity discussed in Section 1.1. To specify the scope of a given performance indicator, it may be necessary to focus on fragments of this process. The example illustrated in the following figure specifies a control for a particular product (500222).

DEFINING THE PERFORMANCE INDICATORS			
Code	01.01.01	Process	Demand Planning and Forecasting 02-feb

1.SPECIFY SCOPE	Process owners:
Define control to report the performance of the forecasting accuracy for new item 500222 to be launched in three demand regions (southern, central, northern). The product will be traded exclusively through distributors. Product 500222 will replace product 500221, which is now commercialized in the central and north regions.	Dem. Planning Manager: T. Miros **Team members:** S.J. Thomas, A. Xavier, T. Yian

Figure 2.2 Specify scope

The scope definition statement indicates the following:

- The need to define a control to report the performance of the forecasting accuracy for new item 500222.
- Item 500222 is to be launched in three demand regions (southern, central, northern).
- The product will be commercialized exclusively through distributors.
- Product 500222 will replace product 500221, which is only commercialized in the central and northern regions.

2.2 Observe Performance

Before defining the specifics of a metric, it is highly recommended that the process is closely observed. If the process is not yet in place, the observation exercise may be done with a similar process.

Given that item 500222 is a new product (that will replace 500221) and it is to be launched in three demand regions (southern, central, northern), and given that 500221 has never been commercialized in the southern region, the process observation should aggregate a set of information related to these premises.

The following figures present the historical demand pattern for product 500221, which is to be replaced. This pattern only shows demand data for the northern and central regions, because it has never been introduced to the southern demand area.

Based on the historical demand, item 500221's forecasting performance has been measured with an indicator named MAPE (mean average performance error). Specifics of MAPE dynamics are detailed in the following chapters.

The table that summarizes MAPE performance in the last fiscal year for item 500221 indicates the northern region faces a higher forecasting inaccuracy profile than the central region. This may suggest that different targets for MAPE according to the demand regions are recommended.

To complete the process observing phase, the reader will find a forecasted demand chart projected for item 500222 is to be launched. Note this chart presents a profile for the southern region, eventually based on experts' inputs or on other similar products (different from 500221) that have been commercialized in this demand area.

This set of information is simply an example, not necessarily conclusive. The more the organization invests in time to generate detailed and complementary information, the more precise the performance will be.

Special attention should be addressed to the graph's details. For example, the following figure illustrates the same set of results in different ways. They represent the monthly record of inventory accuracy in an MRO[1] storage facility. The leftmost graph suggests the inventory accuracy is stable over the last months, and the second graph indicates the process is becoming worse each month.

Note these are illustrations for the same process, same period, same algorithm, and same control mechanism. In fact, the only difference is the y-axis scale.

[1] MRO = type of material. MRO stands for maintenance, repair, operations.

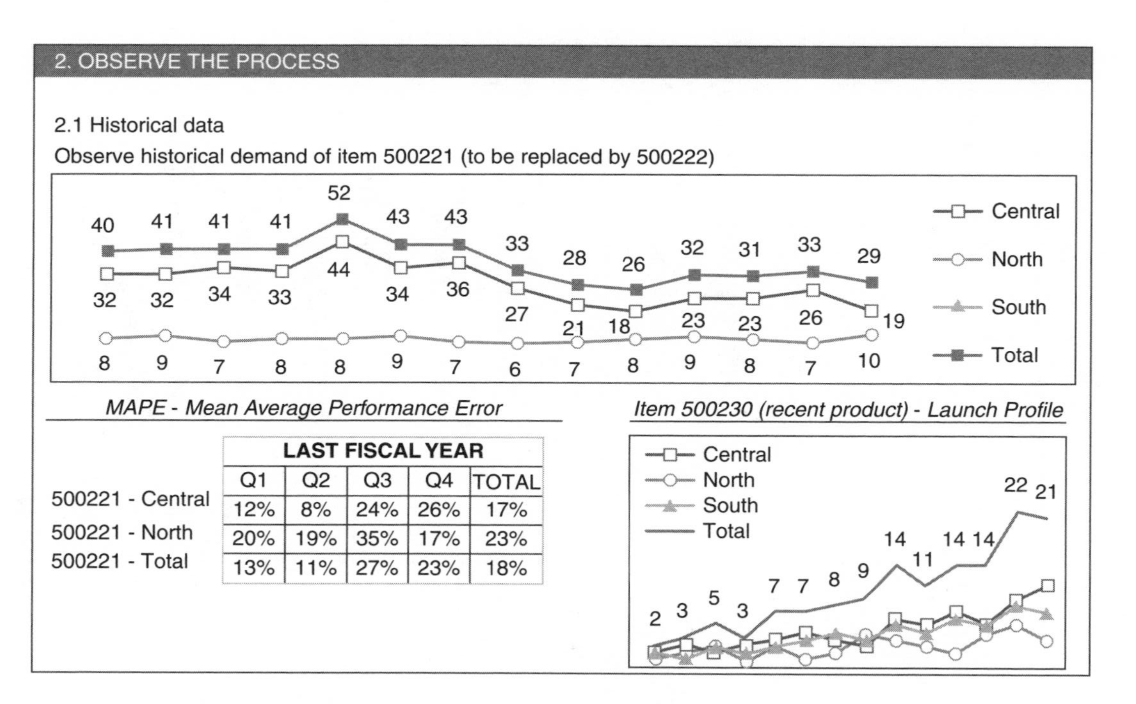

	LAST FISCAL YEAR				
	Q1	Q2	Q3	Q4	TOTAL
500221 - Central	12%	8%	24%	26%	17%
500221 - North	20%	19%	35%	17%	23%
500221 - Total	13%	11%	27%	23%	18%

Figure 2.3 Observe performance

	Jan	Feb	Mar	Apr	May	Jun
Inventory Accuracy	97.10%	96.95%	96.65%	96.20%	95.70%	95.50%

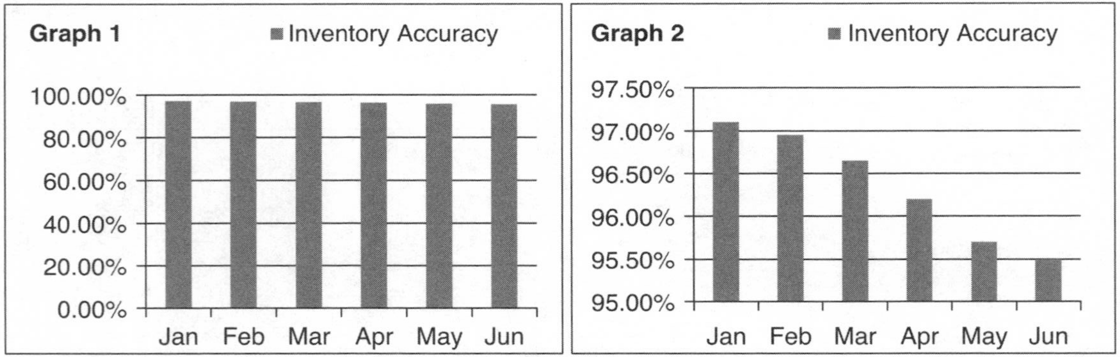

Figure 2.4 The effect of the scales

2.3 Define Performance Pattern

The definition of a target for a given metric frequently requires the elaboration of a pattern that reflects the performance improvement over time, as the organization runs the process learning curve. This performance pattern has as many intermediary stages as needed to show adherence to expected real-world improvement gains.

Our example considers two phases: product introduction and post-introduction phases. Note that targets are more detailed and more aggressive after the introduction phase.

Introduction phase:

- The first 8 months after product launch
- Data captured by demand region
- Reports on a weekly basis
- Forecast error reported as aggregate measure

Post-introduction phase:

- From the ninth month after product launch
- Data captured by demand region
- Reports on a monthly basis
- Forecast error reported by demand region

3. DEFINE PERFORMANCE PATTERN	
3.1 Introduction phase:	Defined to be the first 8 months after product launch Data will be captured by demand region The forecast error will be reported as aggregate measure Report will occur on a weekly basis Error expected to be 25% (+/–5%): 20% to 30%
3.2 Post-introduction phase:	Data will be captured by demand region The forecast error will be reported by demand region Report will occur on a monthly basis Error (Central Region) expected to be 18% to 25% Error (Northern and Southern Regions) expected to be 20% to 27% Error (Total, weighted) expected to be 19% to 24%

Figure 2.5 Performance pattern

2.4 Assign Name for the Control

Despite the apparent simplicity of this step, it is quite important because the name given to the control is "the message" communicated to the rest of the organization. This should be as specific as possible to allow the correct interpretation of the correlation it has with the process it is expected to represent.

If the process performance is to be measured by complementary metrics, and therefore uses different calculation mechanisms (algorithms), these algorithms (and respective metrics) should receive different names.

For example, given the forecast accuracy for product 500222 is to be calculated from the forecast accuracy registered for each demand region (southern, northern, central), it is then necessary to specify how the aggregated metrics will be calculated.

3. DEFINE PERFORMANCE PATTERN	
3.1 Introduction phase:	Defined to be the first 8 months after product launch Data will be captured by demand region The forecast error will be reported as aggregate measure Report will occur on a weekly basis Error expected to be 25% (+/–5%): 20% to 30%
3.2 Post-introduction phase:	Data will be captured by demand region The forecast error will be reported by demand region Report will occur on a monthly basis Error (Central Region) expected to be 18% to 25% Error (Northern and Southern Regions) expected to be 20% to 27% Error (Total, weighted) expected to be 19% to 24%
4. ASSIGN NAME TO THE CONTROL	
4.1 Forecast accuracy error, product 500222, average 4.2 Forecast accuracy error, product 500222, weighted 4.3 Forecast accuracy error, product 500222, central region 4.4 Forecast accuracy error, product 500222, southern region 4.5 Forecast accuracy error, product 500222, northern region	

Figure 2.6 Assign name to the control

This may occur by a simple average calculation or by weighting the final average by the demand quantity in each sales region. In the last case, the forecast accuracy in the higher volume demand region will impact more significantly on the aggregated indicator.

If the organization is not yet comfortable about which algorithm will better fit its control needs, then both metrics should be considered, as illustrated in the previous figure. As the understanding of the process becomes more mature, the organization will decide to keep one of the metrics (usually the weighted average) and discharge the metrics that no longer add value to the business.

It is common that a more complete set of metrics is used in the early stages of a process (introduction phase) while fewer indicators are used to monitor more mature processes.

2.5 Define Algorithm

This step is critical to the success of the governance. A crystal-clear calculation mechanism is fundamental to allow people to properly interpret the process health through indicators' outputs.

The algorithm must be detailed as a sequence of calculations and then illustrated by numbers. Note that no reliable indicator is built upon a qualitative approach. Only a quantitative approach, either through absolute or relative numbers, delivers information that will drive an accurate decision-making process.

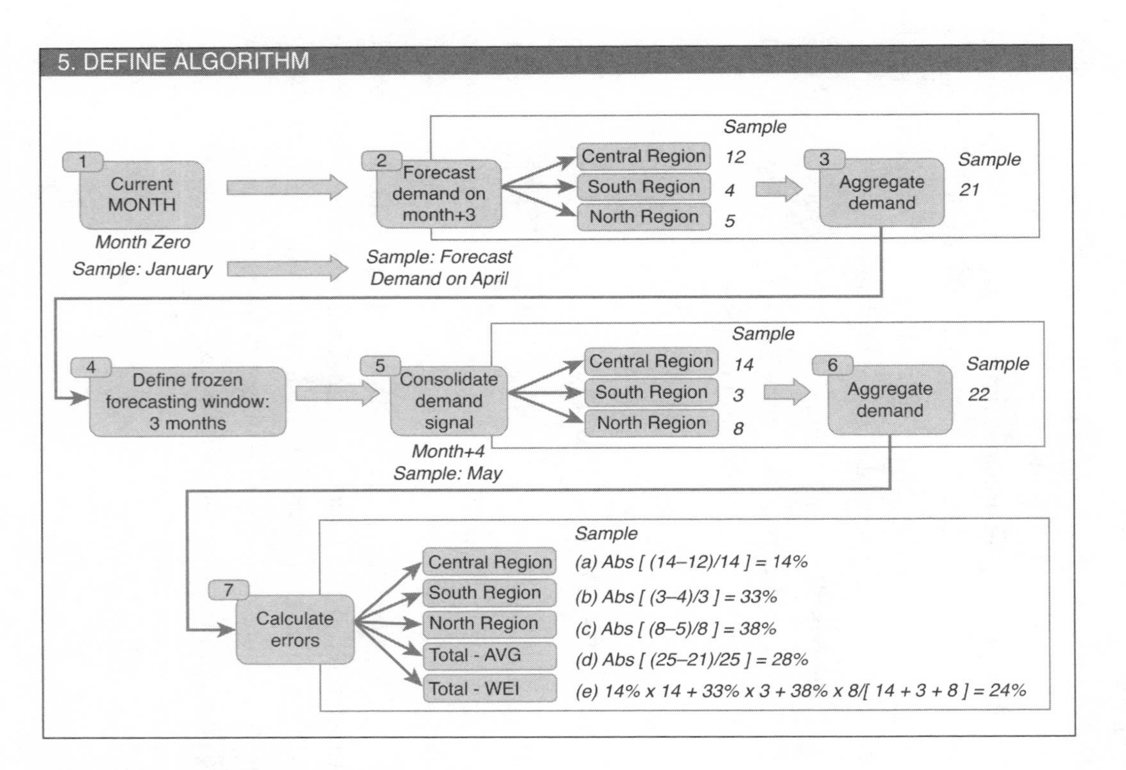

Figure 2.7 Defining metrics define algorithm

2.6 Test Algorithm

The sequence the algorithm follows must be auditable by the organization. Real data has to be treated in a parallel environment—usually in spreadsheets—and reproduce the outputs delivered by the performance indicators' algorithm.

If the algorithm is to be applied to an item that has not yet been released and therefore no real data is available, the audit should be done on any other active item. The example illustrates the testing on active item 200221 with the calculation sequence created for the future new item 500222.

6. TEST ALGORITHM

Routine testing

Central			
FCT	DS		
12	14	Error	Q1
			12
		14%	14
			FY
			12
		14%	14

South			
FCT	DS		
4	3	Error	Q1
			4
		–33%	3
			FY
			4
		33%	3

North			
FCT	DS		
5	8	Error	Q1
			5
		38%	8
			5
		38%	8

Total			
FCT	DS		
21	25	Error	Q1
			21
		16%	25
			21
	WEI	24%	25
	AGV	28%	

Testing on 200221

Central			
FCT	DS		
32	41	Error	Q1
32	30		98
34	40	12%	111
33	41		Q2
44	40		111
34	40	8%	121
36	41		Q3
27	30		84
21	40	24%	111
18	25		Q4
23	30		64
23	31	26%	86
			FY
			357
		17%	429

North			
FCT	DS		
8	9	Error	Q1
9	11		24
7	10	20%	30
8	9		Q2
8	12		25
9	10	19%	31
7	10		Q3
6	11		20
7	10	35%	31
8	9		Q4
9	11		25
8	10	17%	30
			FY
			94
		23%	122

Total			
FCT	DS		
40	50	Error	Q1
41	41		122
41	50	13%	141
41	50		Q2
52	52		136
43	50	11%	152
43	51		Q3
33	41		104
28	50	27%	142
26	34		Q4
32	41		89
31	41	23%	116
			451
	WEI	18.1%	551
	AGV	19.9%	

Figure 2.8 Test the algorithm

2.7 Define Performance Goals

As the calculation algorithm is built and tested and the expected performance patterns have been defined, it is possible to establish the performance quantitative goals, preferably over the time.

This step is recommended to be aligned to the names assigned to the controls. Five different controls have previously been identified for this process:

1. Forecast accuracy error, product 500222, average

   For the first six months: from 22% to 30%

   From the seventh month on: upper limit reduced to 27%

2. Forecast accuracy error, product 500222, weighted

   For the first six months: from 20% to 30%

   From the seventh month on: from 19% to 24%

3. Forecast accuracy error, product 500222, central region

   For the first six months: will not be reported

   From the seventh month on: from 18% to 25%

4. Forecast accuracy error, product 500222, southern region

   For the first six months: will not be reported

   From the seventh month on: from 20% to 27%

5. Forecast accuracy error, product 500222, northern region

   For the first six months: will not be reported

   From the seventh month on: from 20% to 27%

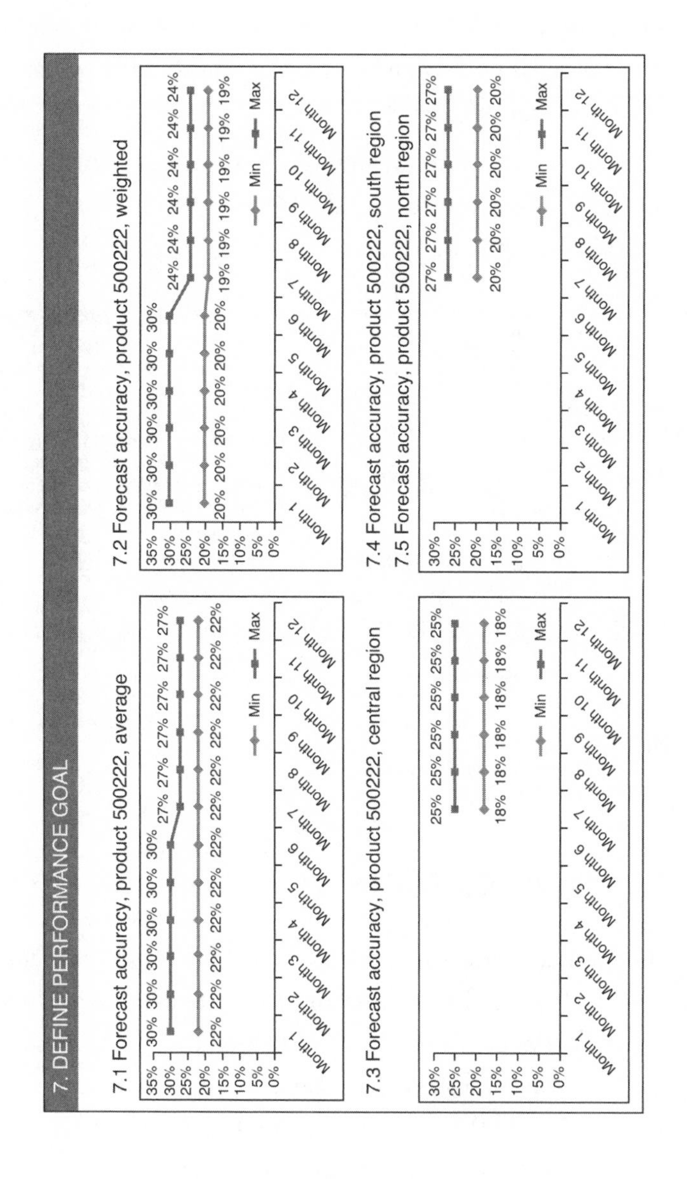

Figure 2.9 Performance goals

2.8 Understand the Learning Curve

Any performance indicator expected to report a changing partner over time requires an explanation. As mentioned before, it is acceptable to expect process performance variation as the organization progresses within the learning curve.

In the example we are following, this pattern has two phases, as described before; these are the product introduction (phase 1) and the post-introduction phase.

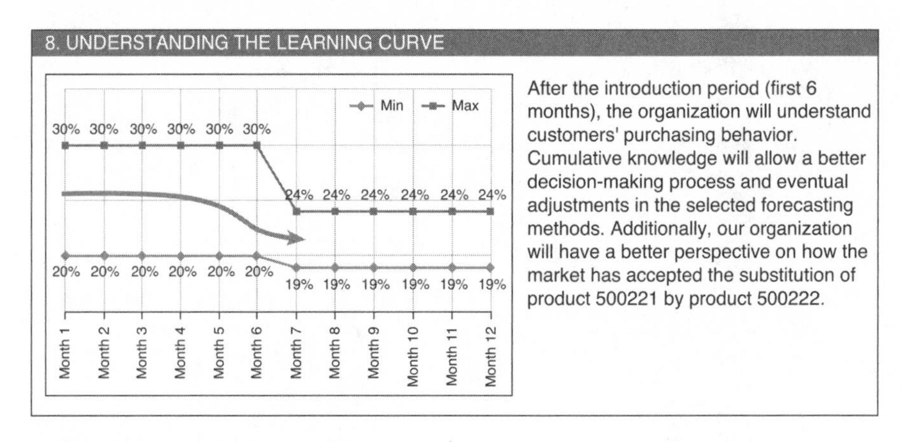

Figure 2.10 The learning curve

2.9 Map Stakeholders

All process stakeholders have to be identified before implementing the new controls and reporting the initial results. Performance indicators do affect people's behavior in relation to their commitment to the process and to other people. The Supply Network Alignment Reference Model (SNAR Model) is a useful tool to support mapping stakeholders of the given metrics used to report a specific process performance. By evaluating each knowledge area identified in the SNAR Model, these can be classified into four categories:

1. Important stakeholders, highly impacted on by the process/metrics
2. Relevant stakeholders, somehow impacted on by the process/metrics
3. Minor stakeholders, with reduced impact caused by the process/metrics
4. Not a stakeholder

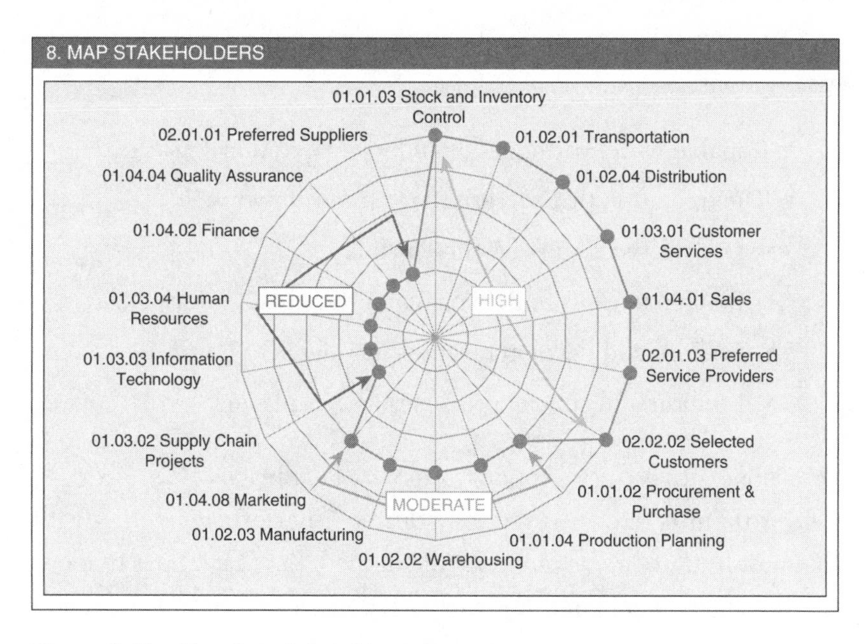

Figure 2.11 Mapping stakeholders

It is no surprise that all important and significant stakeholders should be involved in the next step: validation of the controls' dynamics.

2.9.a Case

This case refers to a study done at a third-party logistics company due to a service-level performance crisis with its major customer. The diagnosis indicated the need to review the business model to realign operational resources and people skills to a new customer service level expectation.

To make matters worse, this new service level had already been contracted, although the supplier had not been delivering it for the previous 4 months. The customer ordered changes to deliver service levels within 2 months or the contract would be canceled and the operation transferred to another logistics service supplier.

The future business model designed with the supplier had these assumptions:

- Capable of delivering stakeholders expectations
- The operation that serves the major customer will:
  - Deliver the results on-time in-full
  - Be able to capture the voice of internal and external clients
  - Acquire and develop business intelligence
  - Monitor and control processes and its players

Though these assumptions seem to be obvious, the effort to deliver them is intense, especially during a performance crisis with the most important customer. As mentioned, the first step is to identify the stakeholders and their expectations. Eight groups of stakeholders were identified:

- Customer (the business)
- Customer's director, who approved the service contract
- Customer's manager, who operates the contract
- Supplier (the business in crisis)
- Supplier's director
- Hauler, hired by the logistics operator

- Regional dealers, responsible for the field operations for the supplier
- Competitors (alternative to actual supplier)

Once the stakeholders are identified, the next step is to understand their wishes, desires, and expectations. How may they influence the process? How can they benefit either from this contract success or failure? These expectations are illustrated in the next figure.

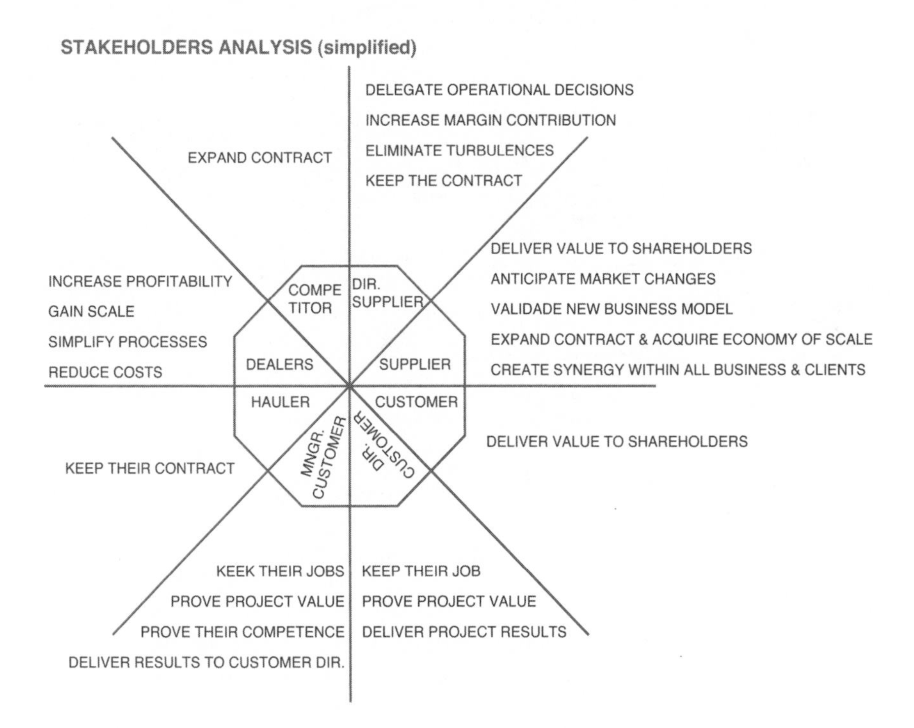

Figure 2.12 Stakeholder expectations

2.10 Validate Control Dynamics

The following table shows how to approach key stakeholders. Each of them should be addressed, either individually or in groups, to assess their commitment to the process and the performance indicator.

9. VALIDATE CONTROL DYNAMICS

STAKEHOLDER AREA		REPRESENTATIVE	COMMITMENT		DO NEXT
Stock and Inventory Control	⇧	F. Mirowsk	✓	High	
Transportation	⇧	J. Guidos	✓	High	
Distribution	⇧	R. Dortman	✓	High	
Customer Services	⇧	W. Wilimann	✓	High	
Sales	⇧	M. Tobas	✗	Low	Schedule new meeting
Preferred Service Providers	⇧	M. Tobas	✓	High	
Selected Customers	⇧	W. Wilimann	✓	High	
Procurement & Purchase	⇨	M. Rockenbach	!	Moderate	Schedule new meeting
Production Planning	⇨	A. Sanders	✓	High	
Warehousing	⇨	A. Munizt	✓	High	
Manufacturing	⇨	A. Sanders	✓	High	
Marketing	⇨	L. Vivix	!	Moderate	Schedule new meeting
Supply Chain Projects	⇩	V. Caius	!	Moderate	Schedule new meeting
Information Technology	⇩	M. Moses	!	Moderate	Schedule new meeting
Human Resources	⇩	D. Spider	✓	High	
Finance	⇩	A. Gims	✓	High	
Quality Assurance	⇩	D. Svantius	✓	High	
Preferred Suppliers	⇩	M. Rockenbach	!	Moderate	Schedule new meeting

Figure 2.13 Validate control dynamics

2.11 Assign Control to People

To close the loop, the defined metrics have to be assigned to specific people. In fact, the mechanism is a little more complex than that. The organization has to identify the sequence from data collection to performance indicator report and make sure that each step is assigned to the right people. This sequence may include the following activities:

- Data collection
- Errors calculation
- Preliminary analysis
- Internal presentation
- Review gaps
- Forum presentation

10. ASSIGN CONTROL TO PEOPLE		
Data collection	S.J. Thomas	Consolidate by 5th each month
Errors calculation	S.J. Thomas	By 7th each month
Preliminary analysis	S.J. Thomas/T. Yian	By 7th each month
Internal presentation	S.J. Thomas/T. Yian	To T. Miros, by 8th each month
Review gaps	S.J. Thomas/T. Yian	By 10th each month
Forum presentation	T.M./S.J.T./T.Y.	To stakeholders, by 12th each month

Figure 2.14 Assign control to people

Each activity has to be ruled by a due date; eventually, this may even define the targeted reporting group. A more detailed approach will define the format of the forms the reports will have.

2.12 From Implementation to Fine-Tuning

The sequence does not finish as the control is implemented. In fact, as this happens, it becomes a very sensitive moment: to monitor the adherence of the control to the process. The performance pattern has to be observed and the algorithm is evaluated and for the first time exposed to real-world dynamics. Eventually, fine-tuning may be required.

Depending on how the business environment reacts to this initial period of implementation, one or more attributes may call for some fine-tuning too. These attributes include the following:

- Adjust algorithm.
- Adjust learning curve profile.
- Adjust stakeholder map.
- Adjust report format.
- Adjust report frequency.
- Adjust people responsibilities.

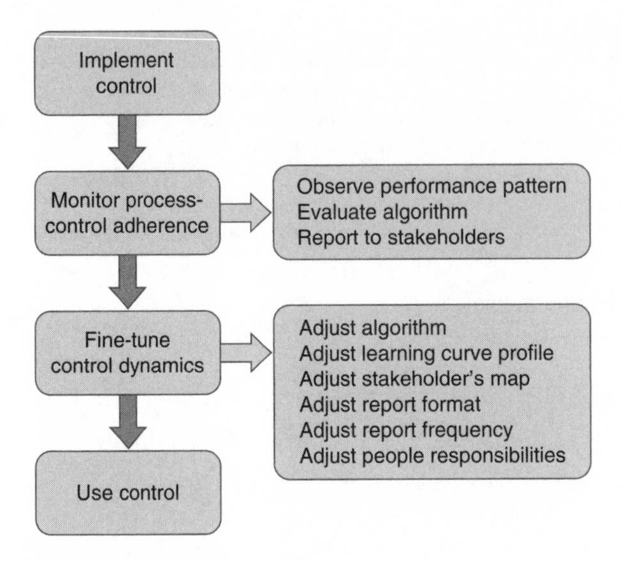

Figure 2.15 Implementation

This moment also requires the first reports to stakeholders. These reporting sessions are strategic to the acceptance of the controls and the success of the whole implementation sequence. Eventual resistance must be minimized at once.

2.13 Reviewing the Cycle

Ideally, the set of controls should be reviewed twice a year. Due to other organizational activities, this reviewing frequency may not occur, but it is mandatory that this effort occurs at least once a year.

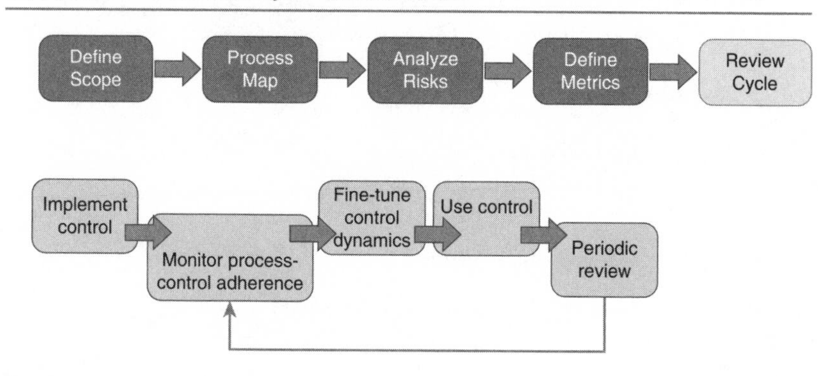

Figure 2.16 Review cycle

2.14 Definition of Metrics in a Project

The definition of performance indicators in supply chain projects requires a different approach. The deployment from project objectives to performance indicators should result from a multifunctional debate with representatives of all stakeholders.

To exemplify this process, we present the set of metrics defined for a CPFR[2] project between a fast-moving consumer goods (FMCG) industry and its key account retailer. The project occurred in the United Kingdom in the early 2000s.

2.14.a Goals to Objectives

The first step is to state the project's *goals* and detail their main *objectives*. These objectives must be adherent to the project's stakeholders. Although this exercise is seemingly intangible, its quality is crucial to the success of the project because it determines the final shape of the project performance metrics.

[2] Collaborative planning, forecasting, and replenishment

Project goal: To improve the collaborative environment to promote the creation of a demand-based supplier order

Stakeholders' objectives:

- Positive impact on customer
- Positive impact on retailer
- Positive impact on supplier
- Identify how to manage push system
- Shape IT needs
- Identify incremental and step changes
- Identify development needs

2.14.b Objectives to Main Activities

The next step is to identify which activities are the enablers for the defined objectives. In some cases, instead of defining an activity, the project team can assign a more specific objective. For example, to have a positive impact on the customer, the project team may wish "to have the product available," instead of stating the key activity is "to sell." This situation is usually resolved in the next step, when the suggested metric will be "monitor sales level" or "monitor revenue value." Each of the seven stakeholders' objectives stated is detailed here:

1. **Positive impact on customer**

   To have the product available

   To have the product at a better price

2. **Positive impact on retailer**

   To have the product available

   Ensure process reliability

   Ensure low-stock policy

   Stock and replenishment easily managed

3. **Positive impact on supplier**

   Have the product available

   Ensure process predictability

   Ensure low-stock policy

   No peak demand periods

4. **Identify how to manage push system**

   Compile and evaluate learning from pilot project

5. **Shape IT needs**

   Compile and evaluate learning from pilot project

   Evaluate technology assessment

6. **Identify incremental and step changes**

   Compile and evaluate learning from pilot project

   Assess capability

7. **Identify development needs**

   Compile and evaluate learning from pilot project

   Assess capability

2.14.c Define the Controls

By defining the controls, the team creates the set of metrics that will evaluate the performance of the operation before, during, and after the project. These metrics can be either external to the project (before or after) or internal (during).

If the project team expects to have an improvement in the overall inventory level, it will hardly be measured during the project. The company will probably assess the inventory level before the project and then, a few days, weeks, or even months later.

Another situation can involve the development of a new hauler to attend a given customer in a new region. During the development, the project team will monitor the performance of more than one hauler to evaluate which one will be selected and contracted. This is an example when internal metrics are determinant.

In this section, the controls are indicated if they are used before the project (pre), during the project, or after the project (post).

Positive impact on the customer

To have the product available:

- *Monitor sales level (pre/post)*
- *Monitor revenue value (pre/post)*

To have the product at a better price:

- *Evaluate cost-benefits to the supply chain (pre/post)*

Positive impact on the retailer

To have the product available:

- *Monitor out-of-stocks level (pre/during/post)*

Ensure process reliability:

- *Monitor hauler delivery accuracy (order and time) (during)*
- *Efficient unloading time (pre/during/post)*
- *Picking on time at customer's distribution center (pre/post)*
- *Retailer distribution center unloading on time (pre/during)*

Ensure low-stock policy:

- *Monitor stock level (average, max, safety stock) (pre/post)*
- *Monitor delivery frequency (during/post)*

Stock and replenishment easily managed:

- *Guarantee more receiver-friendly shipments (during/post)*
- *Monitor shipments size (during/post)*
- *Monitor product damage level (pre/during/post)*

Positive impact on the supplier

Have the product available:

- *Monitor order cycle time (pre/during/post)*

Ensure process predictability:

- *Monitor sales forecast accuracy (pre/post)*
- *Monitor order forecast accuracy (during)*
- *Monitor order cycle accuracy (during)*

Ensure low-stock policy:

- *Monitor stock level (average, max, safety stock) (pre/post)*
- *No peak demand periods*
- *Monitor volume variation (pre/during/post)*

Identify how to manage push system

Compile and evaluate learning from pilot project:

- *Monitor order base arriving on time from supplier (pre/during)*
- *Monitor fixed booking accuracy (during)*

Shape IT needs

Compile and evaluate learning from pilot project:

- *Evaluate technology assessment*
- *Run technology assessment before the project and review after the project (pre/post)*
- *Plan quality "no touch" order (post)*

Identify incremental and step changes:

- *Compile and evaluate learning from pilot project (post)*

Assess capability:

- *Run capability assessment before and after the pilot (pre/post)*
- *Identify the system's needs to set up ongoing solution (pre)*

Identify development needs:

- *Compile and evaluate learning from pilot project*

2.14.d Close the Loop

After the controls are defined, the project team will follow the steps according to the methodology described earlier in this chapter:

- Observe performance, if the project is long enough to allow it.
- Define performance pattern.
- Assign name to the control; eventual adjustment to the activity described in Section 2.14.c.
- Define algorithm.
- Test algorithm.
- Define performance goal.
- Understand learning curve.
- Map stakeholders; already done at Section 2.14.a.
- Validate control dynamics.
- Assign control to people.
- Implement control.
- Monitor process-control adherence, if the project is long enough to allow it.
- Fine-tune control dynamics, if the project is long enough to allow it.

Case: Project Communication

The ability to communicate performance is especially important during projects. The communication channels must be defined by the project team in the early planning stages.

This case illustrates a format to report the project presented in Section 2.14. Note that the report was sent from the project controller to the project leader on a weekly basis; it was titled Weekly Progress Report.

==

PAGE 1: PROGRESS REPORT, WEEK 3

Subject: Ship on Demand Trial – Supplier (Su) / Retailer (Re)

The THIRD "progress report" for the Re/Su trial on "ship on demand" is set out below. The format is

A* comments

B* trial effectiveness measures

- Service level "Su to Re" and "Re DC to Store" / including the use of safety/committed stock)
- Demand amplification analysis
- Re DC stock analysis
- Daily shipping pattern

C* Trial efficiency measures

- Process tracking
  - rounding stock
  - forecast accuracy

D* ABCD review

==

PAGE 2: PROGRESS REPORT, WEEK 3

Subject: Ship on Demand Trial - Su/Re

Progress Report. Week 3 / Overall Results

First 3 weeks (July 10th, July 17th, and July 24th)

A* Comments:

Prior to the trial: The flows of the ten selected product lines (from Su DC to Re DC) were monitored for 4 weeks. The data collected was used to simulate the trial scenario. This simulation helped to define a set of reliable expectations for the trial.

Week 1

By the end of the first week of the Ship on Demand Trial (w/c July 10th), the team confirmed the positive set of expectations for the trial. The initial week of the trial was marked by the set up and fine-tuning of the operational processes.

Week 2

By the end of the second week, the operation ran as planned, although the team has identified room for improvement in the weekend order-handling process.

Week 3

Sound results were achieved in the third week, as the operational processes implemented by the end of week 1 gained momentum and as the weekend order-handling process improvement needs identified in week 2 were successfully implemented.

==

PAGE 3: PROGRESS REPORT, WEEK 3

A. TRIAL EFFECTIVENESS

1. Service level Supplier to Retailer's DC and Re-DC to Store

Results	week 3	Accumulated 10/7 from 17/7
a. Out-of-stocks from Supplier.DC to Retailer.DC	nil	nil
b. Out-of-stocks from Retailer.DC to Store	nil	nil
c. Out-of-stock in Store	nil	nil
d Out-of-stock in Store (2- weeks test)	nil	nil
e. Use of committed stock (# orders)	0	8*
f. Use of committed stock (# orders/total orders)	0%	6%

*(gap = in week 2)

There were no availability issues in the stores. The improvement in the weekend order-handling process enabled a smooth physical (and information) flow. Therefore, the committed stock was not used this week. The team expects the flow pattern achieved in week 3 will be sustained until the end of the trial.

==

PAGE 4: PROGRESS REPORT, WEEK 3

2. Demand Amplification Analysis (cases)

Results	week 2	week 3	accumulated 17/7 24/7
Demand Data	1103	1113	2216
Retailer.DC to Store	1093	1067	2159

variation vs. Sales	–0,.9%	–4,.2%	–-2,.6%
Supplier.DC to Retailer.DC	1022	1208	2230
variation vs. Sales	–7.3%	+8.5%	+0.7%

The volume picked to stores (h) is very similar to the number of cases sold (g) for the same period. The demand amplification between these flows is minimal.

A weekly-basis analysis shows a significant amount of variation between the number of cases shipped from Supplier to Retailer (I) and the number of cases sold, although the variation for the period is very small. The team has identified the causes of these variations and expects a smoother flow from week 4.

==

PAGE 5: PROGRESS REPORT, WEEK 3

In week 2, two factors caused the –7.3% variation.

In the beginning of week 2, there were no orders for several product lines while the initial inventory level was being adjusted. The team allowed the stock to be consumed until each product had only one full-pallet of stock. (This is the agreed size of the committed stock.)

At the end of week 2, there was no Saturday order (for delivery on Sunday).

Therefore, the flow from Supplier to Retailer was 7.3% smaller than the sales volume.

To compensate for the second factor (no order on Saturday), there was a higher shipment size on Monday and Tuesday of week 2. This explains the +8.5% variation.

With the inventory levels adjusted and having the weekend order-handling process improved, the team projects a smooth week 4, with low variation within the physical flows.

==

PAGE 6: PROGRESS REPORT, WEEK 3

3. Stock Level (at Supplier's DC)

Results	week 2	combined 17/7 from 10/7
j. Average (cases/sku)	14.9	17.9
k. Average (% of pallet)	20%	24%
l. Average (equiv. hours of demand)	20 hrs	24.9 hrs

As the weekend order-handling process was improved, the team has avoided the use of committed stock at Retailer's DC.

Therefore, in comparison to the previous week, the overall stock level has increased. Nevertheless, the figures are now closer to the results expected by the team (30 hours of demand) based on the simulation done prior to the trial, which represents a 82% inventory reduction (versus current 7 days of stock) at Retailer's DC.

==

PAGE 7: PROGRESS REPORT, WEEK 3

4. Daily Shipping Pattern (standard deviation of the order size)

Results	End of week 2 End of week 3	17/7 24/7
m. Order Size Variation	107%	90%

(as percentage of order size average)

Two effects have influenced the reduction of the order size variation:

- The improvement of the weekend order-handling process (the variation of week 3 was only 72% vs. 107% of week 2).

- The trial has gained critical mass and, therefore, the effects of minor deviations are now disappearing due to the overall volume shipped.

==

PAGE 8: PROGRESS REPORT, WEEK 3

B. TRIAL EFFICIENCY

1. Process Tracking
   Results
   week 3 Accumulated
   24/7

n. Order on time at Customer Service (10am)	100%	100%
o. Order calculated by 11pm	100%	100%
p. Order downloaded at Supplier's DC at 1pm	100%	100%
q. Order picked and loaded by 4pm	100%	100%
r. Haulers arrived at Supplier's DC by 3pm	100%	100%
s. Hauler left Supplier's DC by 5pm	100%	100%
t. Order arrived at Retailer's DC by 10pm	100%	100%
u. Efficient checking unloading time	100%	100%
v. Order picked to stores on time	100%	93%
w. Orders arrived at stores on time	100%	100%
x. Product not damaged	100%	100%

==

PAGE 9: PROGRESS REPORT, WEEK 3

2. Rounding Stock (round order size to a full pallet layer)

Results

Accumulated

24/7

y. Rounding stock (cases) 3.6

z. Avg rounding stock (as % of daily volume) 21%

Data collected prior to the trial projects that the rounding to a full pallet layer might represent up to 33% of the order size. A volume-expansion analysis projects that this figure might drop to 1% for a volume 150 times higher (considering full-pallet orders).

==

PAGE 10: PROGRESS REPORT, WEEK 3

Forecast Accuracy (Accumulated - forecast vs. Actual)

aa. Day+1 (24h)	–33%
ab. Day+2 (24-48H)	–42%

The SYSTEM report is forecasting figures lower than the actual. (This is the reason for the negative symbol.) This behavior was expected by the team once the data collected prior to the trial projected a forecast accuracy of -18% (Day+1) and -19% (day+2).

Though current figures (–33% and –42%) seem higher than the expected results, the team already expected this pattern at this point of the trial. According to the simulation done prior to the trial, the accumulated forecast accuracy for (day+1) and (day+2) are both high and different from each other in the initial weeks, but as the trial gains critical mass (and the total volume increases), these figures smooth and (day+1) and (day+2) forecast accuracy convergence down to –20%.

Although a plus/minus variation happens on a daily basis analysis, in the long run the CR report consistently predicts less than the actual, across all product lines. This pattern was identified by the team prior to the trial, and the order-calculation process was designed to compensate for this factor. Another team is already working to understand the reasons for this behavior.

======================================

PAGE 11: PROGRESS REPORT, WEEK 3

ABCD

> Achievements

Week 1

SYSTEM Forecast setup done.
New product codes were properly set up.

Week 2

Stores have updated their system with new codes.
All processes worked properly from Monday to Friday.

Week 3

Weekend order-handling process improved Hauler arrival time fixed for 2100hrs (Monday–Friday) and 2200hrs (Sunday).

Order calculation backup process at Newcastle was activated (Thursday and Friday) and sustained the good performance of the beginning of the week.

No use of committed stock.

Process tracking with 100% compliance for all measures!

==

PAGE 12: PROGRESS REPORT, WEEK 3

ABCD

> Benefits

Week 1

A reliable order process is set up.

Data from all stores is available for the SYSTEM forecast.

Trial design is completed.

Week 2

The trial is running as planned.

Week 3

Communication between Retailer.HQ and Suppliers's Customer Service was smooth on Saturday.

Physical flow stability throughout the week.

Information flow stability throughout the week.

Elimination of Hauler waiting time at Retailer's DC (avoiding shift change at 2200hrs) from Monday to Friday (benefit to Hauler and to Retailer's DC).

Improved relationship with selected Hauler.

The product is available for picking to stores one hour before (operational benefit to Retailer's DC).

Backup process at Newcastle was proved robust and reliable.

Reliable order-calculation process.

==

PAGE 13: PROGRESS REPORT, WEEK 3

ABCD

> Concerns

Week 1

SYSTEM forecast remote transmission is slow/delayed/fallen-over (on days when external emails failed). Stores seem to have "incomplete" understanding of the trial operational requirements for the initial week.

Week 2

Weekend order calculation/handling process and backup.

Week 3

Weekend handling process backup at Retailer.HQ.

==

PAGE 14: PROGRESS REPORT, WEEK 3

ABCD

> Do next
when
what identified who

1. Evaluate/fix CR Forecast remote week 1 TJ
> done transmission issues
2. Host message was sent to stores
re-enforcing operational requirements week 1 BE
> done during the set up period

3. Evaluate 2100 arrival at Retailer's DC	week 1	PA/AO
> done (instead of 2200) July 26th		
4. Outline Saturday order calculation	week 1	AO
> done and handling process		
5. Define Saturday backup system	week 2	BE/AO
> partially done		
5.1 Define backup for PGT	week 3	BE
> by end of week 3		

The next report will be issued on August 8th.

Example: Customer Service

Now the reader is presented with an example that refers to the Customer Service knowledge area (SNAR 01.05.01). As a key subprocess (KSP), the order entry was selected.

Code	01.03.01	Processes	Customer Services	02-feb

Key subprocesses (KSP):		Process owners:
KSP a	01.02.02.a Order entry	CS Manager: W. Weltern
KSP b		Team members:
KSP c		A. Vitas. G. Tatus
KSP d		
KSP e		

Supply Network Governance Cycle

Not started = 0.00 Started = 0.25 Completed = 1.00

	KSP a	KSP b	KSP c	KSP d	KSP e	Scale
1. Define scope (apply SNAR model)	1.00					1.00
2. Process Map						
2.a Interviews	1.00					1.00
2.b Process observation	1.00					1.00
2.c Data analysis	1.00					1.00
2.d Organize data	1.00					1.00
2.e Elaborate flows	1.00					1.00
3. Understand business needs	1.00					1.00
4. Define Supply network improvement needs						1.00
5. Elaborate strategies						1.00
5.a Define metrics for strategies						1.00
6. Elaborate policies						1.00
6.a Define metrics for policies						1.00
7. Define practices						1.00
7.a Define metrics for practices						1.00
8. Consolidate governance strategy						1.00
9. Implement cycle revision methodology						1.00
TOTAL SCORING	7.0	-	-	-	-	16.0

KSP a	*44%*
KSP b	
KSP c	
KSP d	
KSP e	
Scale	*100%*

Figure 2.17 Case, the control chart

Following the scope definition, the process has to be mapped. To speed up the presentation of this example, we will now consider that the process has been mapped and registered according to the following figure. After the process is mapped, the next step is to establish targets for the three dimensions: operational, tactical, and strategic.

Operational targets:
1. Implement 95% EDI order entry for key accounts (vs. 78% last fiscal year) 2. Implement 85% EDI order entry for top 20% customers (vs. 71% last fiscal year)
Tactical targets:
1. Define order entry policy considering commercial incentives for EDI 2. Increase IT headcount for Supply Chain Department from 01 to 02
Strategic targets:
1. Assign Customer Logistics Manager for Key Accounts (01 mngr.) 2. Assign Customer Logistics Manager for Top 20% customers (except Key Accounts) (01 mngr.)

Figure 2.18 Case, targets

Following the definition of business targets, the sequence moves on to the metrics definition. The example shows a strategy focused on key accounts (KAs) to achieve 99.0% order entry through EDI. It also indicates a more conservative strategy for other top 20 customers for which it is expected that 85% of order entry comes from the EDI mechanism.

Performance indicators:	Target	Last period	YTD	Comments
1. EDI penetration (KA)	99.0% avg	+ 3%	86.0%	Projected FY = 97%
1.a KA01 Customer AB#03	99.0%	+ 9%	90.0%	Projected FY = 99%
1.b KA02 Customer AB#03	99.0%	+ 0%	89.0%	Projected FY = 99.5%
1.c KA03 Customer AB#03	99.0%	+ 5%	70.0%	Projected FY = 99%
1.d KA04 Customer AB#03	99.0%	+ 18%	95.0%	Projected FY = 99.9%
2. EDI penetration (Top 20%)	85.0% avg	+ 3%	78.7%	Projected FY = 97%
2.a Top 5%	85.0%	+ 12%	81.0%	Projected FY = 85%
2.b Top 10%	85.0%	+ 8%	78.0%	Projected FY = 85%
2.c Top 20%	85.0%	+ 5%	77.0%	Projected FY = 85%

Figure 2.19 Case customer service, performance indicators

The control chart (illustrated in the next figure) also registers the contribution of the last period to the accumulated year-to-date (YTD) results.

The EDI penetration for key account 04 (performance indicator 1.d) increased 18 points last month, contributing to an accumulated result of 95%. This means that before last month, the accumulated result was 77%.

Action plans:	Responsible	Due date	Status
1. Schedule meeting with customer AA#1.	M. Kalmann	23/Feb	To be done
2. Schedule meeting with customer AA#3. Involve IT teams.	M. Kalmann	12/Mar	Confirmed. At customer's office
3. Review distribution strategy for the northwest region. Involve transportation team.	P. Walker	31/Mar	Running new bid
4. Discuss promotional stock strategy with customer AA#2. Involve supply planning team.	P. Walker	15/Mar	Meeting confirmed for 5/mar at CS

Figure 2.20 Case, action plans

The control chart also projects the expected result for the indicator at the end of current fiscal year. For indicator (1.d), this projection indicates 99.9% of order entry on key account 04 will be via EDI.

To make sure that performance targets are achieved, some actions are required. These actions may lead to the creation of supporting projects—example (3) in the next figure.

Code	01.03.01	Processes	Customer Service	02-feb

Key subprocesses (KSP):		Process owners:
KSP a	01.02.02.a Order entry	CS Manager: W. Weltern
KSP b		
KSP c		Team members:
KSP d		A. Vitas. G. Tatus
KSP e		

Supply Network Governance Cycle

Not started = 0.00 Started = 0.25 Completed = 1.00

	KSP a	KSP b	KSP c	KSP d	KSP e	Scale
1. Define scope (apply SNAR model)	1.00					1.00
2. Process Map						
2.a Interviews	1.00					1.00
2.b Process observation	1.00					1.00
2.c Data analysis	1.00					1.00
2.d Organize data	1.00					1.00
2.e Elaborate flows	1.00					1.00
3. Understand business needs	1.00					1.00
4. Define Supply network improvement needs						1.00
5. Elaborate strategies						1.00
5.a Define metrics for strategies						1.00
6. Elaborate policies						1.00
6.a Define metrics for policies						1.00
7. Define practices						1.00
7.a Define metrics for practices						1.00
8. Consolidate governance strategy						1.00
9. Implement cycle revision methodology						1.00
TOTAL SCORING	7.0	-	-	#	-	16.0

KSP a	44%
KSP b	
KSP c	
KSP d	
KSP e	
Scale	100%

Operational targets:

1. Implement 95% EDI order entry for key accounts (vs. 78% last fiscal year)
2. Implement 85% EDI order entry for top 20% customers (vs. 71% last fiscal year)

Tactical targets:

1. Define order entry policy considering commercial incentives for EDI
2. Increase IT headcount for Supply Chain Department from 01 to 02

Strategic targets:

1. Assign Customer Logistics Manager for Key Accounts (01 mngr.)
2. Assign Customer Logistics Manager for Top 20% customers (except Key Accounts) (01 mngr.)

Performance indicators:	Target	Last period	YTD	Comments
1. EDI penetration (KA)	99.0% avg	+ 3%	86.0%	Projected FY = 97%
1.a KA01 Customer AB#03	99.0%	+ 9%	90.0%	Projected FY = 99%
1.b KA02 Customer AB#03	99.0%	+ 0%	89.0%	Projected FY = 99.5%
1.c KA03 Customer AB#03	99.0%	+ 5%	70.0%	Projected FY = 99%
1.d KA04 Customer AB#03	99.0%	+ 18%	95.0%	Projected FY = 99.9%
2. EDI penetration (Top 20%)	85.0% avg	+ 3%	78.7%	Projected FY = 97%
2.a Top 5%	85.0%	+ 12%	81.0%	Projected FY = 85%
2.b Top 10%	85.0%	+ 8%	78.0%	Projected FY = 85%
2.c Top 20%	85.0%	+ 5%	77.0%	Projected FY = 85%

Action plans:	Responsible	Due date	Status
1. Schedule meeting with customer AA#1.	M. Kalmann	23/Feb	To be done
2. Schedule meeting with customer AA#3. Involve IT teams.	M. Kalmann	12/Mar	Confirmed. At customer's office
3. Review distribution strategy for the northwest region. Involve transportation team.	P. Walker	31/Mar	Running new bid
4. Discuss promotional stock strategy with customer AA#2. Involve supply planning team.	P. Walker	15/Mar	Meeting confirmed for 5/mar at CS

Figure 2.21
Case, control chart overview

Example: Quality Assurance

This example presents limited comments to allow you to analyze it without additional support.

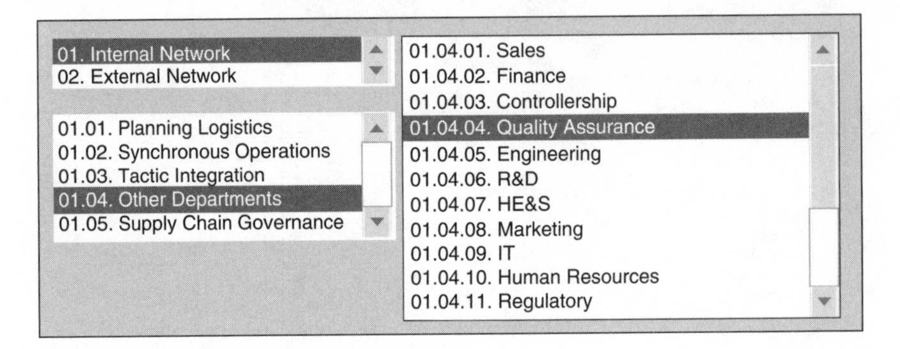

Figure 2.22 Defining the scope

Code	01.04.04	Processes	Quality Assurance	*02-feb*

Key subprocesses (KSP):		Process owners:
KSP a	01.04.04.a Raw material analysis time (chemical lab)	QA Manager: D. Svantius
KSP b	01.04.04.b Raw material analysis time (bio lab)	**Team members:**
KSP c		R. Rivers, P. Vivini
KSP d		
KSP e		

Figure 2.23 Defining key subprocesses

Code	01.04.04	Processes	Quality Assurance	02-feb

Key subprocesses (KSP):		Process owners:
KSP a	01.04.04.a Raw material analysis time (chemical lab)	QA Manager: D. Svantius
KSP b	01.04.04.b Raw material analysis time (bio lab)	
KSP c		Team members:
KSP d		R. Rivers. P. Vivini
KSP e		

Supply Network Governance Cycle

Not started = 0.00 Started = 0.25 Completed = 1.00

	KSP a	KSP b	KSP c	KSP d	KSP e	Scale
1. Define scope (apply SNAR model)	1.00	0.25				1.00
2. Process Map						
2.a Interviews	1.00	0.25				1.00
2.b Process observation	1.00	0.25				1.00
2.c Data analysis	1.00	0.25				1.00
2.d Organize data	1.00	0.25				1.00
2.e Elaborate flows	1.00					1.00
3. Understand business needs	1.00					1.00
4. Define Supply network improvement needs	1.00					1.00
5. Elaborate strategies	1.00					1.00
5.a Define metrics for strategies	1.00					1.00
6. Elaborate policies	1.00					1.00
6.a Define metrics for policies	1.00					1.00
7. Define practices	1.00					1.00
7.a Define metrics for practices	1.00					1.00
8. Consolidate governance strategy	1.00					1.00
9. Implement cycle revision methodology						1.00
TOTAL SCORING	15.0	1.3	-	-	-	16.0

KSP a	94%
KSP b	8%
KSP c	
KSP d	
KSP e	
Scale	100%

Figure 2.24 Case quality assurance, monitoring KSP evolution

Operational targets:
1. Reduce RM chemical analysis average lead time to 02 days (vs 2.7 days last fiscal year) 2. Increase chemical lab productivity to 16 analyses per analyst per day (vs 12 last fiscal year)
Tactical targets:
1. Implement segmented performance indicator for Analysis Cycle Time (ACT) 2. Reduce production stops - less than 1% due to ACT above 01 day (vs 2.2% last fiscal year)
Strategic targets:
1. Enable inventory reduction by 1.2% (vs. last fiscal year scenario)

Figure 2.25 Case quality assurance, targets

Performance indicators:	Target	Last period	YTD	Comments
1. Analysis cycle time (ACT)	<= 02 days	2.7	2.31	Projected FY = 2.0
1.a From RM receiving to sampling	< 0.7	na	1.01	Projected FY = 0.8
1.b From sampling to report (analysis)	< 1.0	na	0.90	Projected FY = 0.9
1.c From report to system release	< 0.3	na	0.40	Projected FY = 0.3
2. Production stops	1.0% ACT>1d	1.5%	1.8%	Projected FY = 1.2

Action plans:	Responsible	Due date	Status
1. Evaluate RM sampling methodology and capacity	R. Rivers	14/Mar	To be done
2. Define correlation between RM ACT and inventory reduction	D. Svantius	12/Mar	Confirmed. At customer's office

Figure 2.26 Case quality assurance, controls and actions

Expert's Opinion

Competencies and Skills for Tomorrow's Supply Chain Managers

Professor Martin Christopher[3]

Supply chain management as an idea is still young. Real organizational implications of managing the end-to-end pipeline are not always fully understood. There is a limited availability of managers with the appropriate skills and capabilities to manage complex global supply chains.

[3] Martin Christopher, Emeritus Professor of Marketing and Logistics at Cranfield University, United Kingdom

What many companies call supply chain management might more accurately be described as logistics or even distribution management. If we adopt the definition of supply chain management that I like to use, it becomes clear that a key role of supply chain management is to focus as much on managing outside the boundaries of the business as it is about internal concerns: The purpose of supply chain management is to manage upstream and downstream relationships with suppliers and customers in order to deliver superior value in the final market at less cost to the supply chain as a whole.

The more that companies move into multichannel marketing and distribution, the greater is the need to work more closely with intermediaries. The implications of managing across boundaries in what some have termed the 'extended enterprise' are profound. Particularly challenging are the implications for the managerial skills and competencies that will be required if the supply chain truly is to be managed end to end.

One critical implication of the adoption of the supply chain management philosophy is for the organizational architecture of the business. For centuries, companies have used an organizational logic based on "division of labor," whereby activities take place within functions or departments. Although this functionally based organizational concept may ensure the efficient use of resources, it is inwardly focused and tends to lead to a silo-like mentality. These types of organizations also tend to lack customer responsiveness and hence are slow to respond to changes in the marketplace.

Put simply, the requirement is to transform the organization from an inwardly-focused "vertical" structure to an outwardly focused "horizontal" business. The horizontal organization has a number of distinguishing characteristics. It is:

- Market facing
- Organized around processes

- Built upon cross-functional teams
- Guided by metrics that are customer-centric

It is the focus on processes rather than functions that distinguishes the horizontal organization. These teams will comprise specialists drawn from the functional areas (which now become "centers of excellence") and will be led by "integrators," whose job it is to focus the process team around the achievement of market-based goals. In such organizations, a different type of skills profile is clearly needed for managers at all levels.

What are the skills and capabilities needed by these integrators? One model that is gaining acceptance is the idea of the T-shaped manager. A T-shaped manager has a specific functional specialization (the down-bar of the *T*) but also has a strong understanding of the different activities that take place across the end-to-end supply chain process (the cross-bar of the *T*).

So, as a hypothetical example, the process team leader might have a background in inventory management, his functional specialism. To be successful as an integrator, that team leader must also have an understanding of all the other activities that are involved in converting an order into cash.

Therefore, they will need to be familiar with the relevant information systems technology, with costing tools such as activity-based costing, and with appropriate planning frameworks such as sales and operations planning (S&OP). In addition, the supply chain integrator needs to be capable of recognizing and managing the sources of complexity in their supply chain. To assist in this task, they will have an understanding of the tools and techniques of Business Process Reengineering and Six Sigma methodologies. To round this off, they will be very familiar with the latest thinking on supplier relationship management and customer relationship management. Quite a challenge!

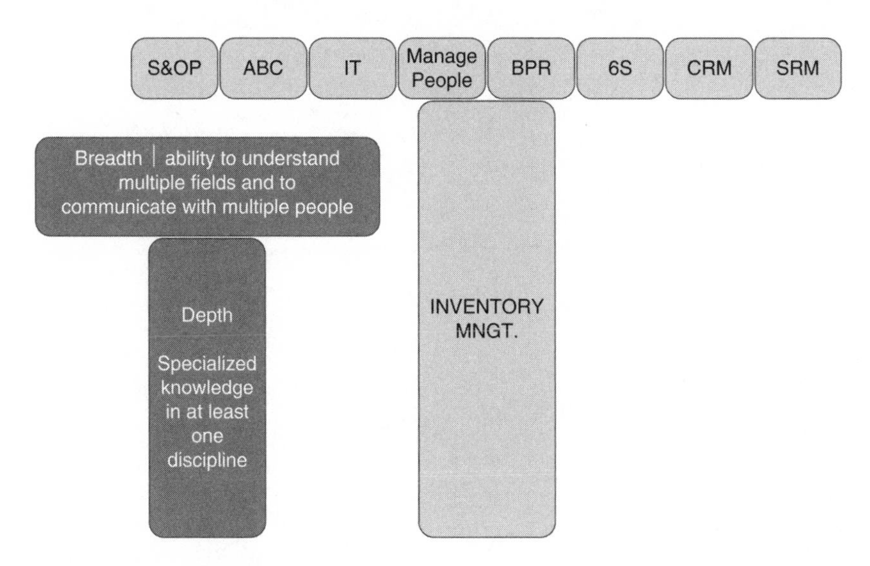

Figure 2.27 T people

It will be clear from this description of the skills and capabilities required to become a successful supply chain integrator that there are probably at this moment few people who could match the profile. The implications of this are clear: The skills and capabilities that are fundamental to the role of supply chain integrator cannot be acquired solely through osmosis and experience but also through appropriate management education programs. Fortunately, a growing number of business schools and other organizations now offer advanced courses covering these subjects.

3

Performance Indicators

Performance indicators are applicable to any area of the business—in fact, to any knowledge area of life. The variety of potential controls requires that categorization methodologies support the process detailed in Section 1.4 in which a metric is defined.

One possible approach to organize metrics related to operations is by using four pillars: time, productivity, quality, and cost. Another approach classifies the performance indicators into three categories: efficiency, efficacy, and adaptability.

Figure 3.1 Performance indicators basic categorization

3.1 Efficiency, Efficacy, Adaptability

Performance indicators that report process efficiency are the most often used because they preserve strong adherence to the concepts of productivity, speed, and quality. Efficiency indicators express the resource utilization, the elimination of waste or obsolescence, and the search for productivity—resource utilization versus resource capacity. Usual metrics capture process cycle times and percentages of resources utilization. Examples include the following:

- Delivery on time (% of delivery on time versus total shipments)
- Damages (% of damaged cases versus total shipped cases)
- Picking quality (% cases picked correctly versus total cases picked)
- Vehicle conditions (% vehicle in good conditions versus total vehicles)

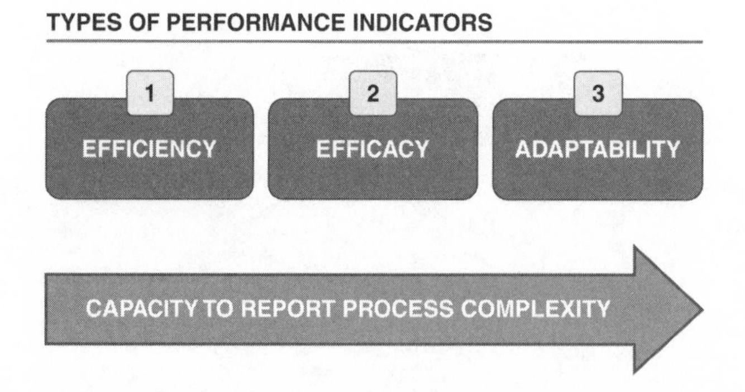

Figure 3.2 Capacity to report process complexity

The mechanisms that measure process efficacy are considered to be either tactical or strategic and may result from a combination of the efficiency indicators.

Efficacy indicators usually illustrate the performance of an end-to-end process (large group of activities) or some of its subprocesses (limited group of activities)—differently from efficiency indicators that usually report the performance of a specific activity.

Metrics considered as efficacy indicators focus on process durability, punctuality, accuracy, costs, repeatability, and quality.

A classic example is the *on-time in-full* (OTIF), which represents the performance of a complex process composed of both subprocesses and activities. Because the OTIF may have different calculation algorithms, the following example illustrates three possible calculation sequences:

- On-time delivery (efficiency indicator) = 97%
- Delivery without damages (efficiency indicator) = 99%
- Perfect invoicing (efficiency indicator) = 98%
- Right quantity delivered (efficiency indicator) = 96%
- Vehicle in good condition (efficiency indicator) = 99%

The first possible algorithm based on the calculation of simple average is OTIF = [97+99+98+96+99] / 5 = 97.8%.

The second possible algorithm is based on the weighted average. Each subprocess is classified according to various possible criteria. This example considers the following weighting:

- On-time delivery = 3
- Delivery without damages = 2
- Perfect invoicing = 3
- Right quantity delivered = 2
- Vehicle in good condition = 1

Therefore, the calculation sequence is OTIF = [(3 × 97) + (2 × 99) + (3 × 98) + (2 × 96) + (1 × 99)] / [3 + 2 + 3 + 2 + 1] = 97.6%.

The third algorithm uses the multiplication of each component: OTIF = 97% × 99% × 98% × 96% × 99% = 89.4%.

OTIF - On Time in Full			
Alternative Algorithms			
Simple Average	**Weighted Average**		**Multiplication**
97.0%	3	2.91	97.0%
99.0%	2	1.98	99.0%
98.0%	3	2.94	98.0%
96.0%	2	1.92	96.0%
99.0%	1	0.99	99.0%
	11	10.74	
97.8%		97.6%	89.4%

Figure 3.3 Alternative algorithms (1)

The difference between the use of these algorithms is perceived as the performance of at least one of the subprocesses varying. Some algorithms may be more sensitive to this variation than the others, as you will notice.

This small 3% chance in one of the subprocesses is reflected in the final OTIF calculation differently. The intensity of change is smaller in the second algorithm (–0.28%) as subprocess (e) was weighted 1.0 while other subprocesses were ranked either 2.0 or 3.0. A small change was also registered for the first algorithm (simple average) with a –0.61% variation for the aggregated OTIF.

The following figure compares the algorithms' sensitivity. The multiplication method presents the highest variation (–3.03%), which is ten times the variation presented for the weighted average mechanism. This attribute (sensitiveness) is highly desirable in a performance indicator.

Although the third mechanism is now regarded as the most adequate, executives are often uncomfortable with it as it reports lower percentages. A typical question is "Why report a 89.4% performance if there is another similar mechanism that reports 97.8%?"

OTIF - On Time in Full			
Alternative Algorithms			
Simple Average	Weighted Average		Multiplication
97.0%	3	2.91	97.0%
99.0%	2	1.98	99.0%
98.0%	3	2.94	98.0%
96.0%	2	1.92	96.0%
99.0%	1	0.99	99.0%
	11	10.74	
97.8%		97.6%	89.4%

OTIF - On Time in Full				
Alternative Algorithms				
Simple Average		Weighted Average		Multiplication
97.0%		3	2.91	97.0%
99.0%		2	1.98	99.0%
98.0%		3	2.94	98.0%
96.0%		2	1.92	96.0%
96.0%	-3.0%	1	0.96	96.0%
		11	10.71	
97.2%			97.4%	86.7%
-0.61%			-0.28%	-3.03%

Figure 3.4 Alternative algorithms (2)

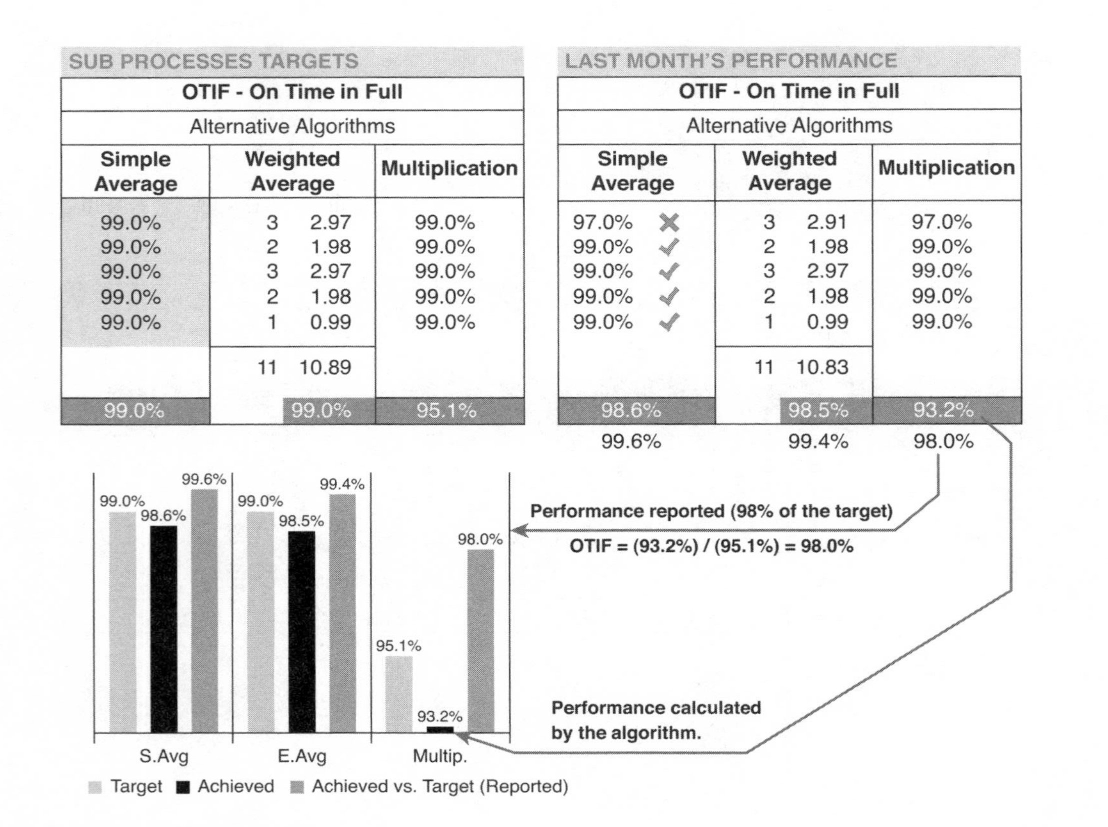

Figure 3.5 Adjusting the report

The correct approach requires a target-based adjustment on how to report the performance of OTIF using the multiplication process. The following example simulates the impact of a 97% performance of process (a), on-time delivery, when all subprocesses have the same target of 99%.

As expected, the multiplication algorithm shows the higher oscillation (from targeted 95.1% to achieved 93.2%). If the organization is to report the output of the algorithm's calculation (93.2%), it may communicate the wrong message. Therefore, an adjustment is conducted.

The achieved result is compared to the target. This will accurately inform that the OTIF performed 98% of the expected target, which is more likely to communicate the right message. This adjustment is recommended as usually the maximum value for a given performance indicator is 100%, which is not the case for OTIF based on the multiplication algorithm.

Another classic example of an efficacy performance indicator is the *line fill rate* (LFR), which represents the number of order lines delivered to the client compared to the total lines ordered. Suppose, for example, that the organization delivered 11 out of 15 ordered lines: LFR = (Ordered line attended) / (Total ordered lines) = 11 / 15 = 73%.

The LFR depends on several subprocesses such as supplier and production performance, forecasting accuracy, distribution efficacy, inventory holding policy, and many others. Because the LFR performance indicator output represents the aggregated performance of a set of complex processes, it is considered to be an efficacy indicator.

An efficacy indicator may be built from efficiency indicators or by other efficacy indicators or even by a combination of the two. This mechanism is illustrated in the next figure. The adaptability indicators may also be built according to various combinations of other metrics.

The adaptability performance indicators are usually customized to address the specific needs of the organization.

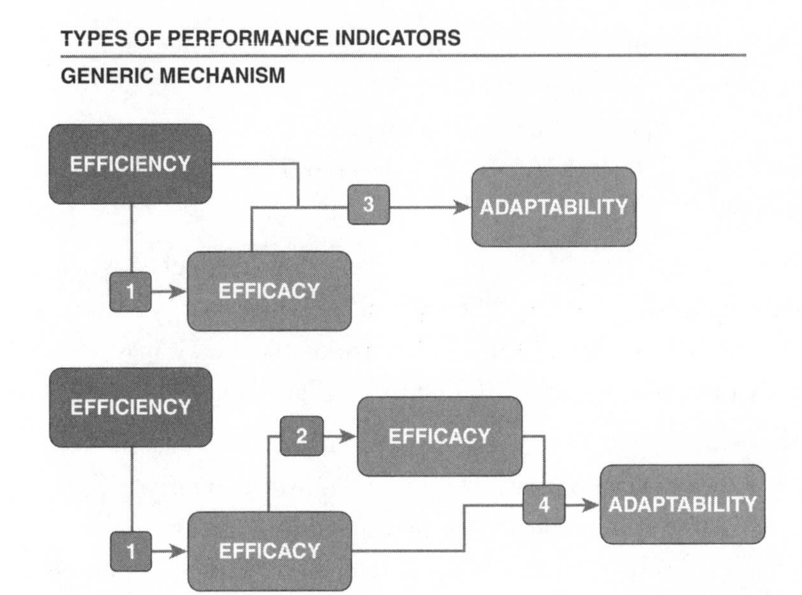

Figure 3.6 Types of performance indicator

Some simple correlations are examples of adaptability performance indicators. A buyer responsible for the cereals category will anticipate purchase orders of beans when weather forecasts indicate lower temperatures.

The correlation between sales of beans and temperature may be structured as shown in the next figure. The leftmost column indicates that when the daily temperature is 2.5°C below the month's average temperature, it is expected that beans sales will be 2.2% above the average sales pattern for this month.

On the opposite side of the graph, when the temperature is 2.5°C above the month's average temperature, it is expected that beans sales will be 2.5% below historical demand.

This correlation was established over time, by cross-checking historical data and eventual qualitative information, which allowed building a unique correlation that is unlikely to be useful to any other business.

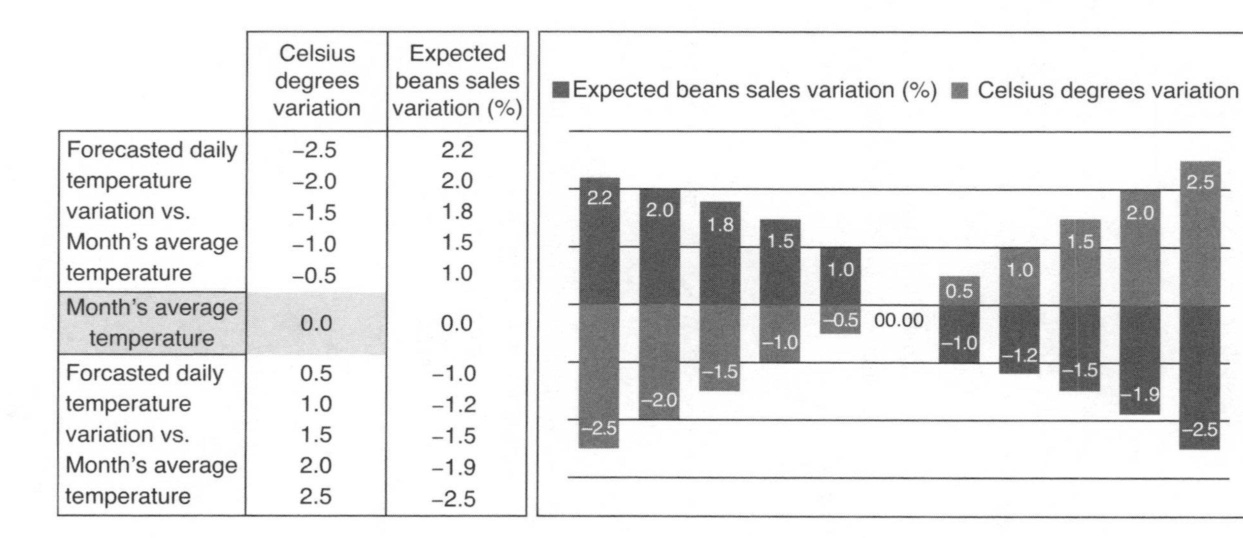

	Celsius degrees variation	Expected beans sales variation (%)
Forecasted daily temperature variation vs. Month's average temperature	–2.5	2.2
	–2.0	2.0
	–1.5	1.8
	–1.0	1.5
	–0.5	1.0
Month's average temperature	0.0	0.0
Forcasted daily temperature variation vs. Month's average temperature	0.5	–1.0
	1.0	–1.2
	1.5	–1.5
	2.0	–1.9
	2.5	–2.5

Figure 3.7 Performance indicator, adaptability

A more sophisticated model may define the correlation between forecasting accuracy and sales revenue—or lost sales. An organization correlates its aggregated finished goods forecast accuracy to the LFR performance. As expected, the higher the forecast accuracy, the higher the LFR.

The correlation captured forecasting accuracy data from 83% to 88% and plotted this against the observed LRF, which oscillated from 92.9% to 99.5%. As indicated in the following figure, if the forecast accuracy increases from 83.5% to 88% (4.5 points), it is expected that the LFR will increase from 94.1% to 99.5% (5.6 points).

The organization then associates LFR performance with sales revenue. At the top LFR performance (99.5%), which is to be achieved if forecast accuracy reaches 88%, the expected sales revenue for the fiscal year is 800 million euros.

But because sales forecasting is less accurate, the LFR reduces and the company loses revenue. When the sales forecast is 8.5%, the LFR reduces to 94.1% and the organization's revenue reduces by 43,420,000 euros (5.4% loss).

This correlation (adaptability performance indicator) suggests that if this organization is capable of sustaining an average 88% forecasting accuracy performance, then the business revenue will increase by 5.4%. Now this organization may consider the demand planning areas strategic to the business!

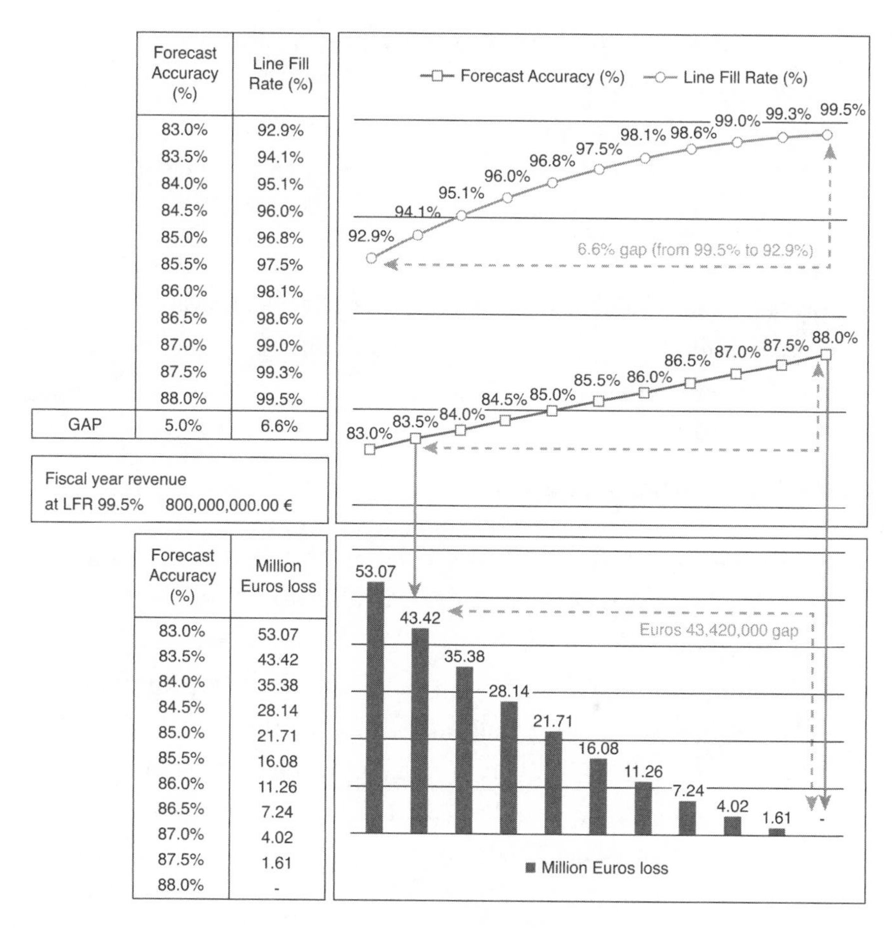

	Forecast Accuracy (%)	Line Fill Rate (%)
	83.0%	92.9%
	83.5%	94.1%
	84.0%	95.1%
	84.5%	96.0%
	85.0%	96.8%
	85.5%	97.5%
	86.0%	98.1%
	86.5%	98.6%
	87.0%	99.0%
	87.5%	99.3%
	88.0%	99.5%
GAP	5.0%	6.6%

Fiscal year revenue at LFR 99.5%	800,000,000.00 €

Forecast Accuracy (%)	Million Euros loss
83.0%	53.07
83.5%	43.42
84.0%	35.38
84.5%	28.14
85.0%	21.71
85.5%	16.08
86.0%	11.26
86.5%	7.24
87.0%	4.02
87.5%	1.61
88.0%	-

Figure 3.8 Performance indicator, sophisticated approach

3.2 Applications

Section 1.1 presented the SNAR Model, illustrated in Figure 1.2, as the best alternative to support scope definition within the Supply Network Governance Cycle (SNG Cycle). In this section, the SNAR Model structure offers a transparent approach to organize several performance indicators applicable to all supply chain knowledge areas (Oliveira and Gimeno 2014_01).

The number of performance indicators is nearly infinite. It is not possible to produce a conclusive list of metrics that covers all supply chain areas. Previous sections have also explained that one control may vary due to adjustments in attributes such as calculation algorithm or frequency of report.

Although you will find in this section at least one example of performance indicators for most of the SNAR Model knowledge areas, it is not intended to be a definitive list. Academics and practitioners are always debating this theme.

3.2.a Planning Logistics

Purchase reliability: This evaluates the adherence of purchase orders' arrival dates versus planned arrival dates. Note the indicator "Raw material (RM) accuracy" is 50%. This is supported by a table that registers raw materials forecasted receipt dates (fourth row). The following row shows data indicating whether this raw material has actually arrived. This is a very simple approach as it basically monitors the "on time"; more sophisticated algorithms may be elaborated, including variables such as quality or documentation expanding the perspective to "on-time in-full." The example is based on data from a large chemical industry headquartered in the United Kingdom.

Expenses versus targets: This simple indicator compares current expenses versus targets. The radar chart (see Figure 3.10) is a powerful mechanism to report multiple indicators simultaneously and in the same unit (%). In fact, this "second layer" is the mechanism to follow the process performance, while the radar chart is an interface tool.

This example is based on data from a large chemical industry headquartered in Germany. The figure illustrates the control this organization has on purchase saving, comparing it to commercial conditions previously defined in valid contracts. This figure also has a few elements that require some explanation.

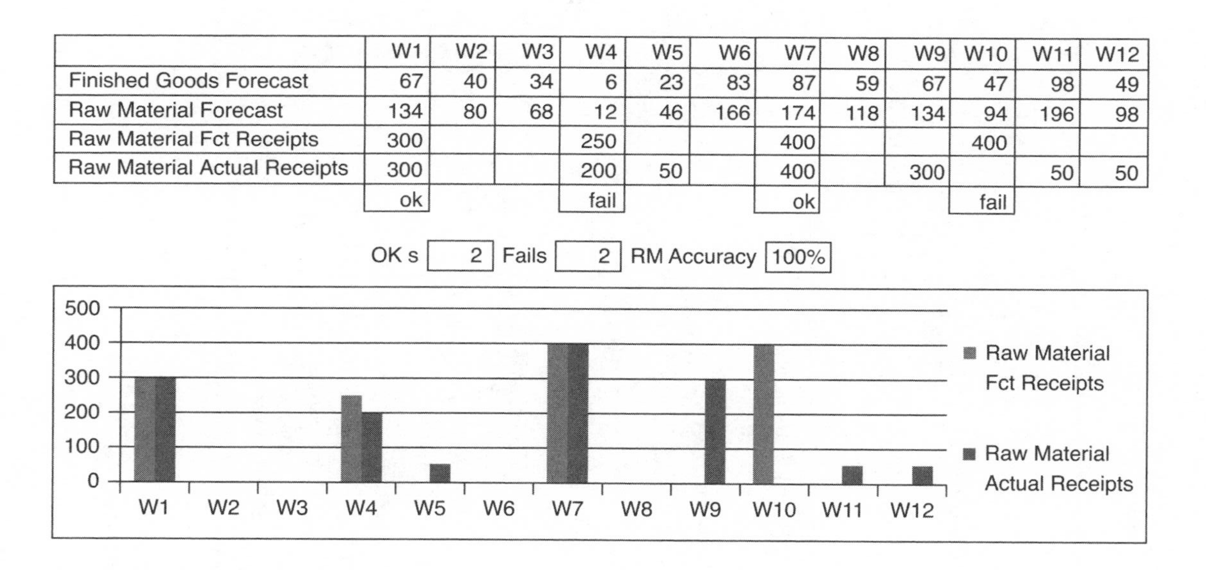

	W1	W2	W3	W4	W5	W6	W7	W8	W9	W10	W11	W12
Finished Goods Forecast	67	40	34	6	23	83	87	59	67	47	98	49
Raw Material Forecast	134	80	68	12	46	166	174	118	134	94	196	98
Raw Material Fct Receipts	300			250			400			400		
Raw Material Actual Receipts	300			200	50		400		300		50	50
	ok			fail			ok			fail		

OK s 2 Fails 2 RM Accuracy 100%

Figure 3.9 Purchase reliability

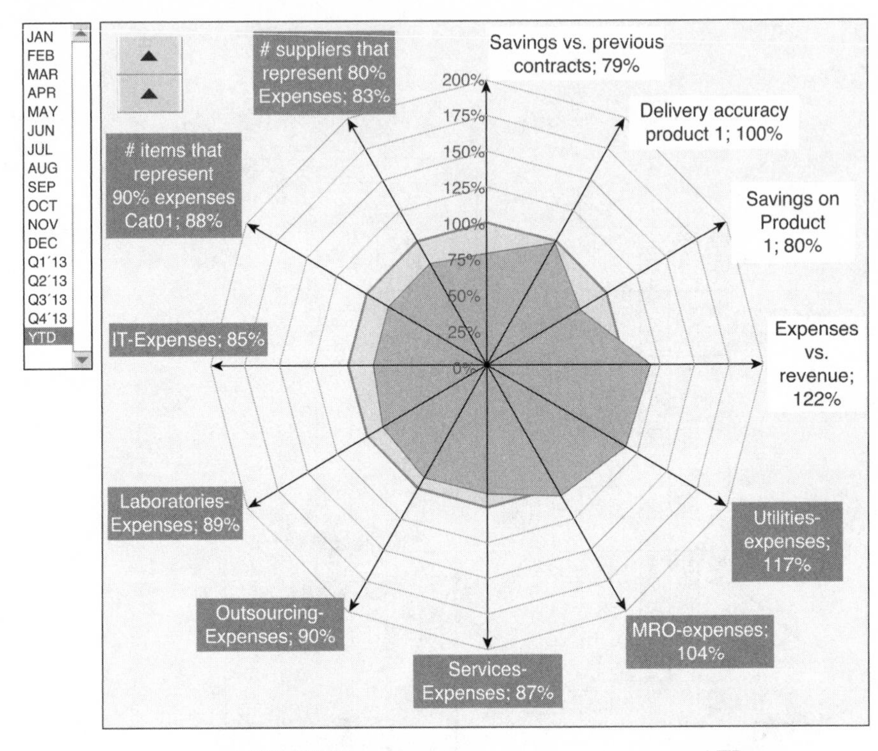

	INDICATORS	YTD
1	Savings vs. previous contracts	79%
2	Delivery accuracy product 1	100%
3	Savings on Product 1	80%
4	Expenses vs. revenue	122%
5	Utilities-expenses	117%
6	MRO-expenses	104%
7	Services-Expenses	87%
8	Outsourcing-Expenses	90%
9	Laboratories-Expenses	89%
10	IT-Expenses	85%
11	# items that represent 90% expenses Cat01	88%
12	# suppliers that represent 80% Expenses	83%

Figure 3.10 Radar chart

The monthly target is $650.00. This company registers the savings and compares them to the target. For example, the savings in May were $605, which is only 93% of the monthly target.

The control also creates comparisons for aggregated periods such as the quarter or year. The accumulated savings for the third quarter (July, August, and September) has a monthly average of $271 ([86 + 230 + 497] / 3 = 271), which is only 42% of the expected target. In the year-to-date (YTD), this organization has saved a monthly average of $516 (21% below the target). In the radar chart, this information is available in the upward vertical axis (first indicator). The same mechanism, reporting the process as percentage of the target, is applied to the other 12 processes illustrated in the radar chart.

Sales forecast accuracy: This is an effective indicator whose goal is to measure the ability of the division (or department) to assess the turnover in the future. It can be measured per item, per family of products, customer, and business. It is suggested that it be measured and evaluated monthly. The calculation of this indicator can be accomplished by the sales ratio provided for in the last 3 months and sales provided for in the last month.

This example is based on data from a large FMCG (food industry) headquartered in Denmark. This company forecasts the next 11 months of demand based on the past 13 months. The algorithm calculates variation around the polynomial curve to evaluate potential forecasting error.

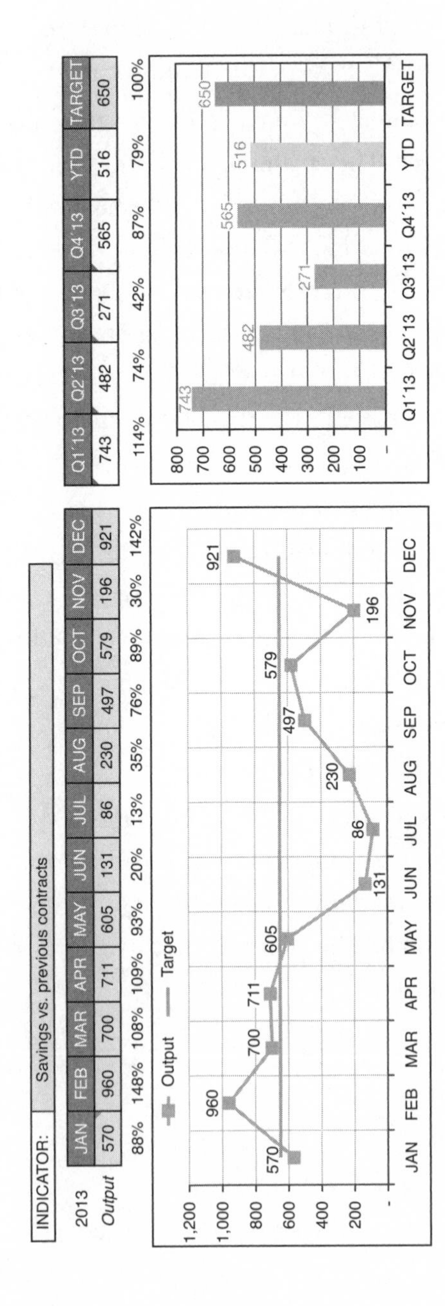

INDICATOR:	Savings vs. previous contracts											
2013	JAN	FEB	MAR	APR	MAY	JUN	JUL	AUG	SEP	OCT	NOV	DEC
Output	570	960	700	711	605	131	86	230	497	579	196	921
	88%	148%	108%	109%	93%	20%	13%	35%	76%	89%	30%	142%

Q1´13	Q2´13	Q3´13	Q4´13	YTD	TARGET
743	482	271	565	516	650
114%	74%	42%	87%	79%	100%

Figure 3.11 Radar chart, component

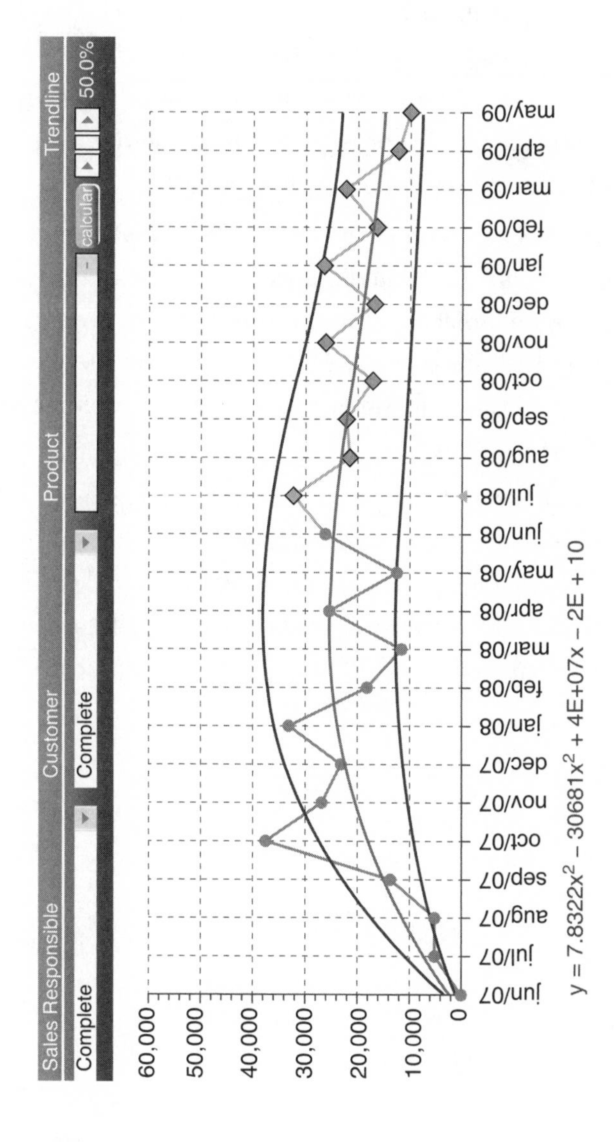

Figure 3.12 Forecasting

Demand and price volatility: Several businesses find finished goods demand to be under the strong influence of pricing policies. Despite the fact that all commodity sectors are known to react to price variation, other segments such as pharmaceutical and chemical follow a similar pattern. The following chart illustrates price variation versus units sold. In this case, the correlation is nonexistent. Sometimes the role of a performance indicator is simply to demystify assumptions that may lead to a wrong decision-making process.

This example is based on data from an FMCG (cosmetics industry) headquartered in the United States and refers to a direct sales channel. There is no direct correlation with price and promotion. The demand is under the influence of fashion and media. There is a "mismatch" between sales and shipments (that is, low capacity to change short-term plans).

MAPE: MAPE stands for "mean absolute percentage error." It divides the absolute deviation by the data (usually the volume in units) to obtain the percentage error. Other complementary algorithms are CFE (cumulative forecast error), MAD (mean absolute deviation), and MSE (mean squared error). The idea is to track forecasting error across time.

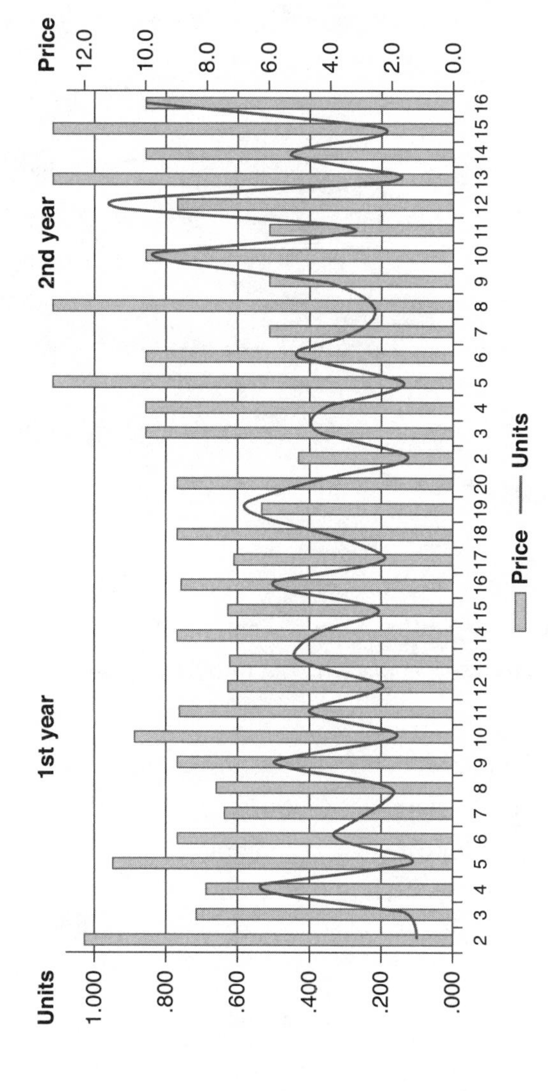

Figure 3.13 Demand versus price variation

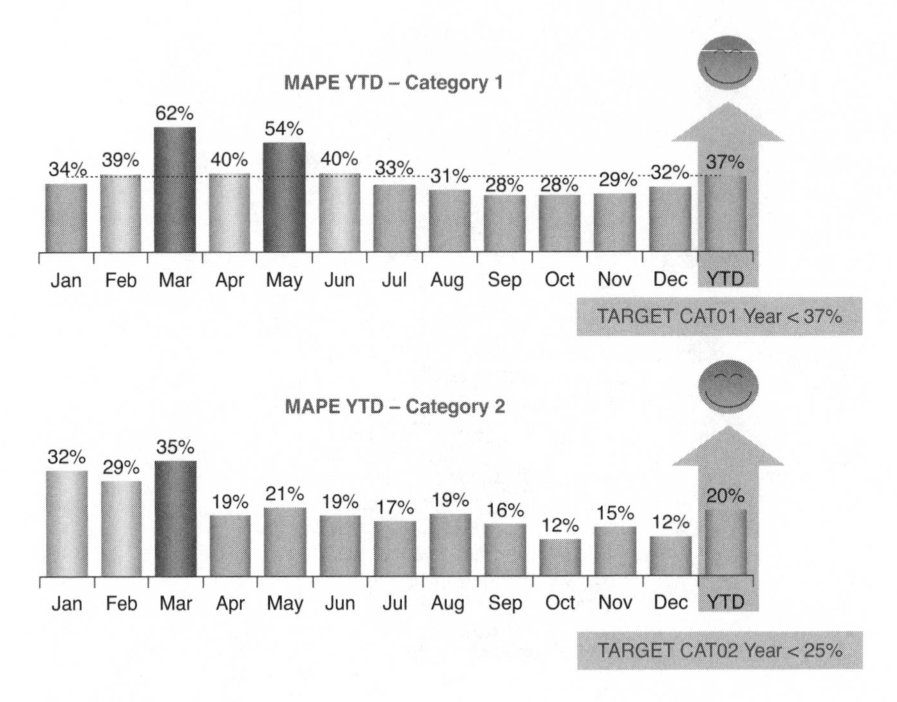

Figure 3.14 MAPE

This example is based on data from an FMCG (food industry) headquartered in the United States. Analysis of the facts behind the graphs allows us to understand why the indicator oscillated in specific months.

January and February suffered from pricing changes. In January, competitors increased their prices, and in February they reduced them. Both graph 2 (for category 1) and graph 3 (for category 2) show an increased MAPE in March as the business entered a new demand region. A similar effect occurred for category 1 in May due to publicity on television.

MAPE versus service level: This is an example of an indicator categorized as having adaptability; Section 2.1 covers this concept. It correlates two indicators of efficacy (MAPE and service level). The service level may represent different scenarios, as explained in Section 1.4, especially when detailing the calculation algorithm.

In theory, whenever the demand forecasting error reduces, product availability, and thus service level, increases. This type of indicator should be used supported by a statistical analysis: the correlation index that is available in any electronic spreadsheet set of functions.

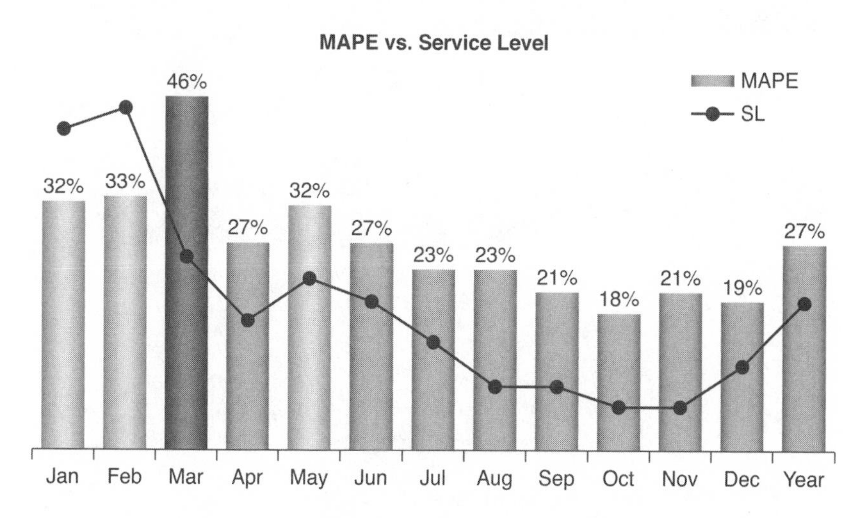

Figure 3.15 MAPE versus service level

MAPE – segmentation: This method, as well as CFE, MAD, and MSE, may be applied according to various segmentation strategies.

Forecasting: The forecasting process offers multiple alternatives in terms of performance indicators. This is a graphical representation of past demand visually connected to demand forecast for the following months. Statistical treatment was used to define the acceptable upper and lower limits for the forecast. This serves to question if the first two points forecasted (and also the last two points) are really to be considered. In fact, the forecasted profile looks pretty unreliable. To offer a mechanism to criticize the process is an important attribute of any performance indicator. This example is based on data from a medium-sized pharmaceutical industry headquartered in Argentina. The trend, based on recent months, is represented by the dotted line upward. UCL is the upper limit, LCL the lower limit.

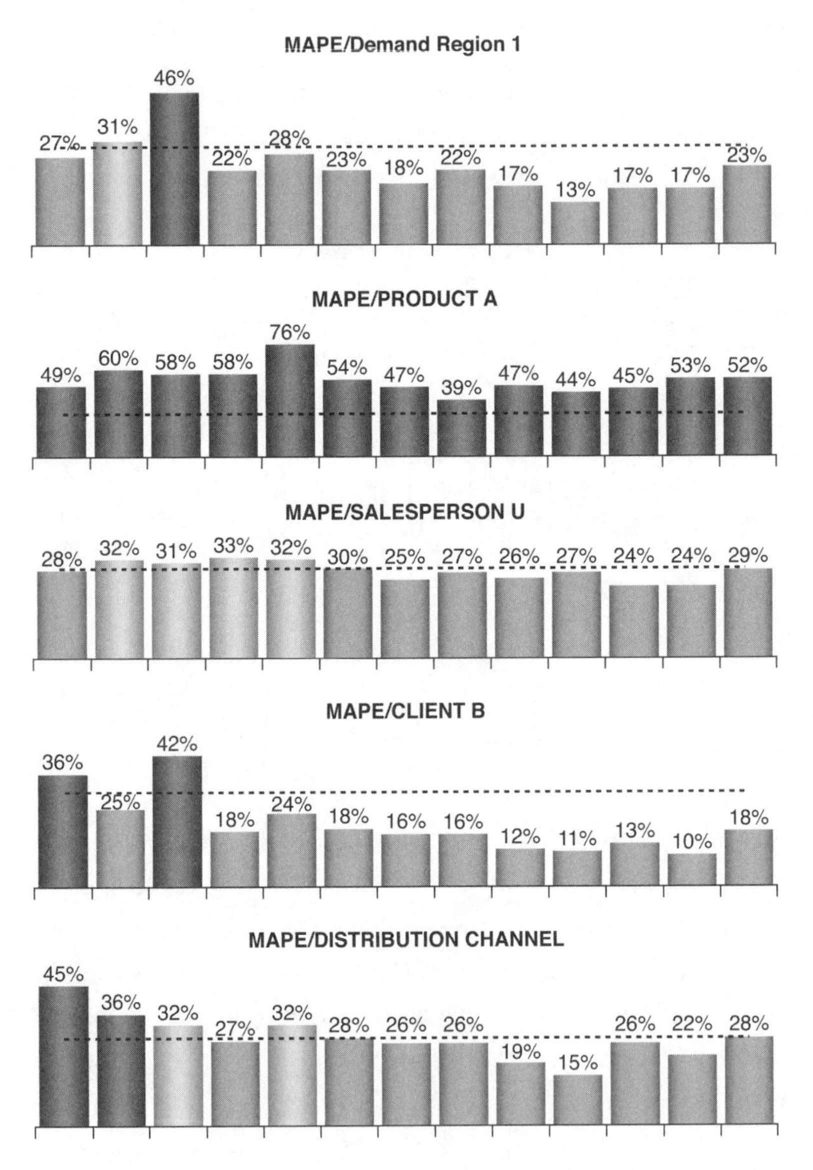

Figure 3.16 MAPE – segmentation

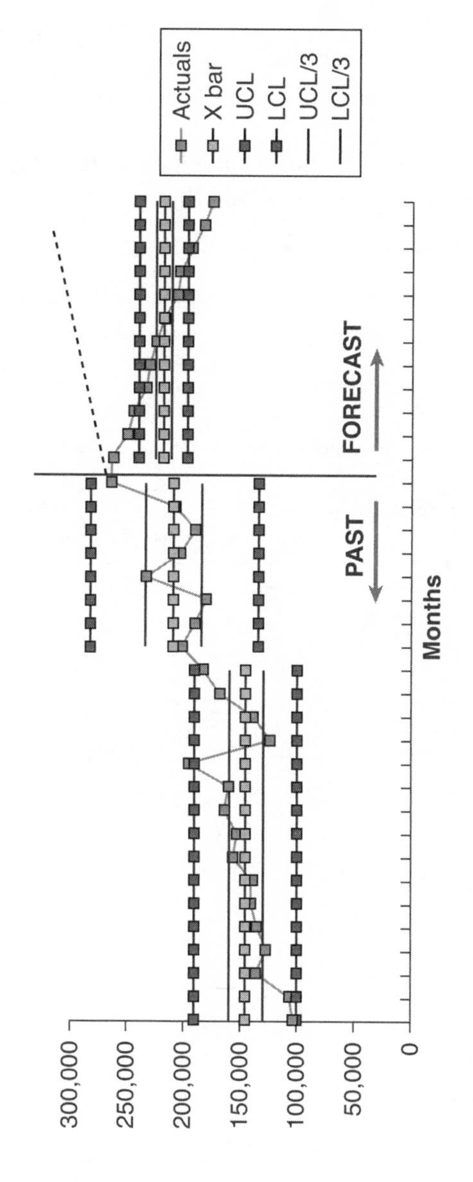

Figure 3.17 Forecasting

Raw materials availability to production: This indicates the efficacy in the process of the acquisition of raw materials. It can be measured by item, family, supplier, or ABC curve. It is calculated as the ratio of the number of occurrences and the number of items produced. The example below presents a "one-sheet" format that combines the indicator report with the continuous improvement mechanism. This example is based on data from a large multinational automotive parts industry headquartered in Italy. Performance reports in the graph are real (bars) versus target (line). It also offers space to register the "problems" that affected the process and thus influenced the performance indicator. It also reserves space to register action plans and their status.

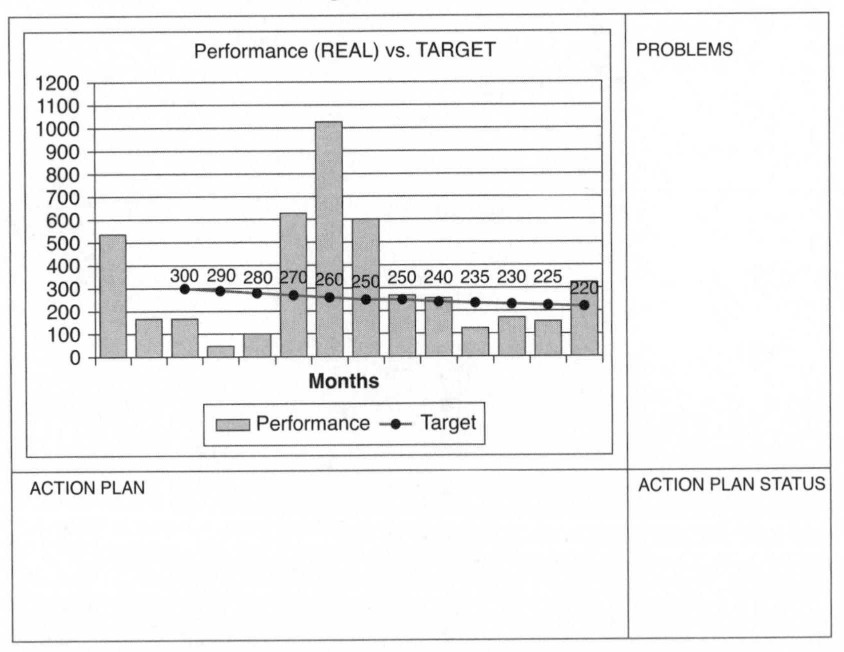

Figure 3.18 Hours lost due to lack of parts

Purchase cycle time: This is measured from the creation of a purchase order until the goods are delivered to the target location, usually a warehousing operation.

Price variation: This control monitors the price variation of specific items (product or service) or those segmented into groups of products. This price variation is compared to inflation in the same period. Ideally the price monitored should not be increasing (positive variation) more than the reported inflation during the period.

This example is based on data from a large public hospital located in Brazil. Indicator type A reports variation within 6 months, and indicator types B and C report variation within 3 months.

Supplier's delay: A supplier's late delivery is a measure of on-time delivery (OTD). For example, two pieces worth $2.00 each totals $4.00. This example is based on data from a large multinational automotive industry headquartered in Germany. There are two categories of supplier: UP and CONV. The values of their delays is summed and represented by the bars. The target is 3.5 (million, local currency).

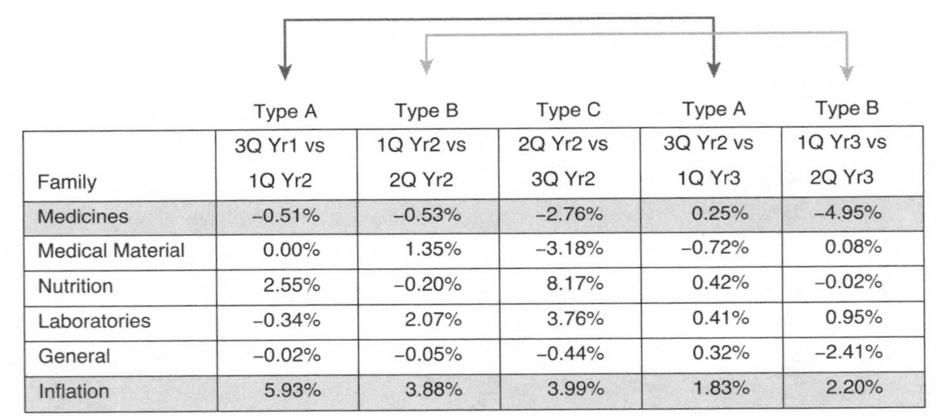

Family	Type A 3Q Yr1 vs 1Q Yr2	Type B 1Q Yr2 vs 2Q Yr2	Type C 2Q Yr2 vs 3Q Yr2	Type A 3Q Yr2 vs 1Q Yr3	Type B 1Q Yr3 vs 2Q Yr3
Medicines	–0.51%	–0.53%	–2.76%	0.25%	–4.95%
Medical Material	0.00%	1.35%	–3.18%	–0.72%	0.08%
Nutrition	2.55%	–0.20%	8.17%	0.42%	–0.02%
Laboratories	–0.34%	2.07%	3.76%	0.41%	0.95%
General	–0.02%	–0.05%	–0.44%	0.32%	–2.41%
Inflation	5.93%	3.88%	3.99%	1.83%	2.20%

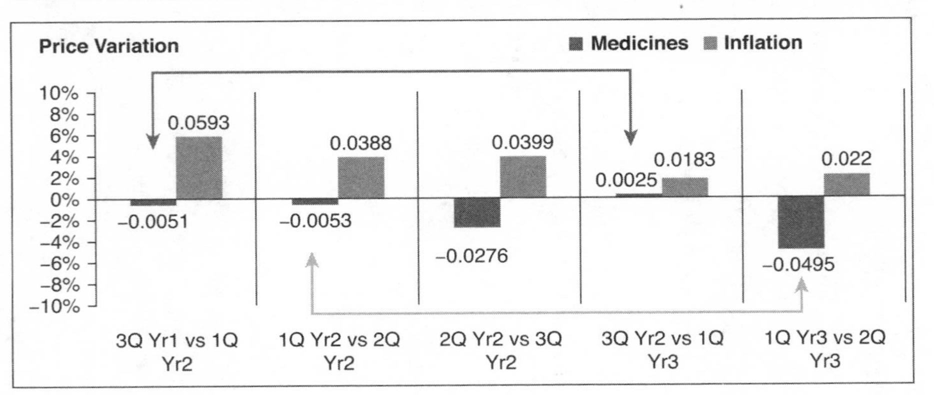

Figure 3.19 Price variation

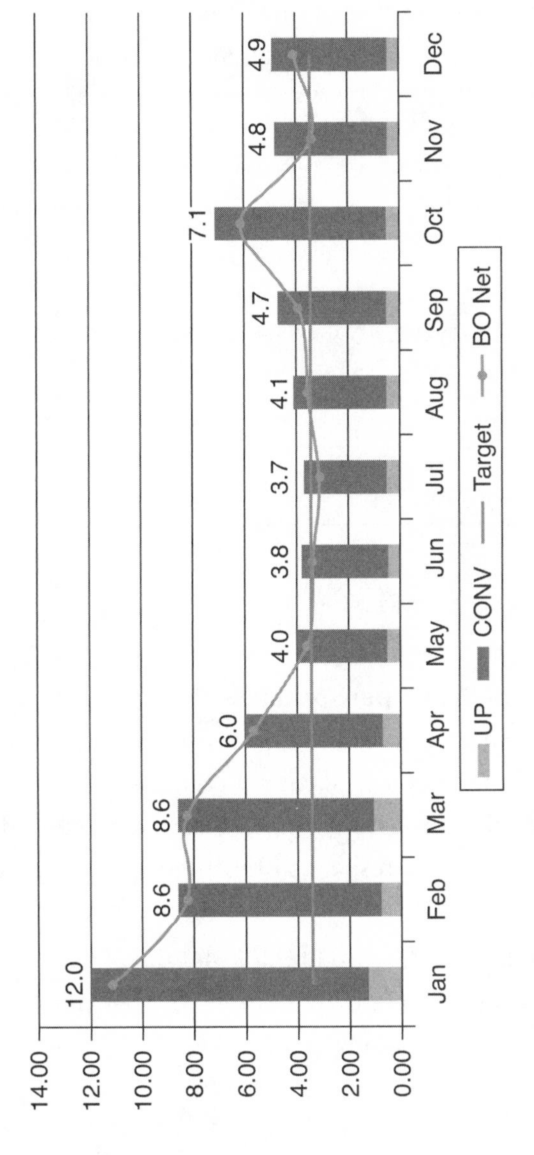

Figure 3.20 Value of supplier's delay

Purchase on-time delivery (OTD): This indicator is equivalent to the previous one (that is, supplier's delay). In the following example, there is a target of 90% for purchase order OTD. The graph shows the performance variation for three categories. The example is based on data from a large multinational aerospace industry headquartered in Brazil.

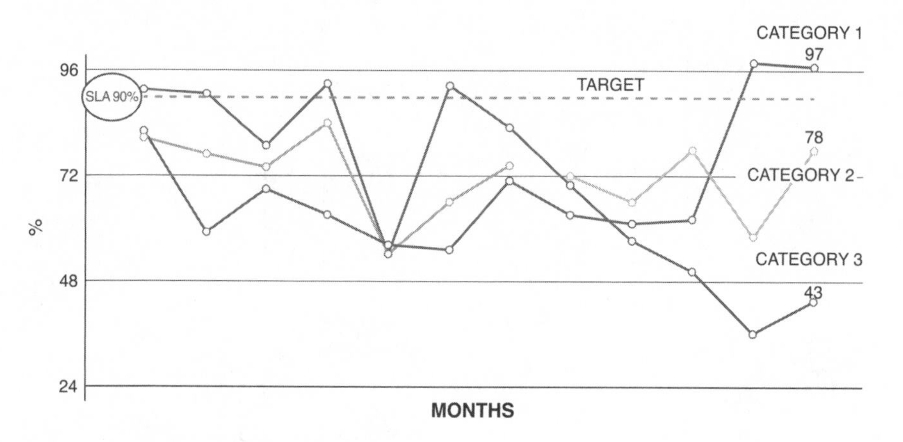

Figure 3.21 Supplier's service level

Supplier's delivery performance (SDP): SDP indicates the performance of the suppliers in delivering orders. It is obtained by the ratio between the number of correct deliveries and the total number of deliveries; the correct deliveries are considered to be those that present compliance with respect to the product, quality, date, and amount requested.

Supplier's service level: This indicates the efficacy of suppliers delivering on time and "in full." The definition of *in full* includes multiple attributes such as the quality of products, absence of damages, correct documentation and others. This offers a different perspective to that of supplier's delay discussed earlier.

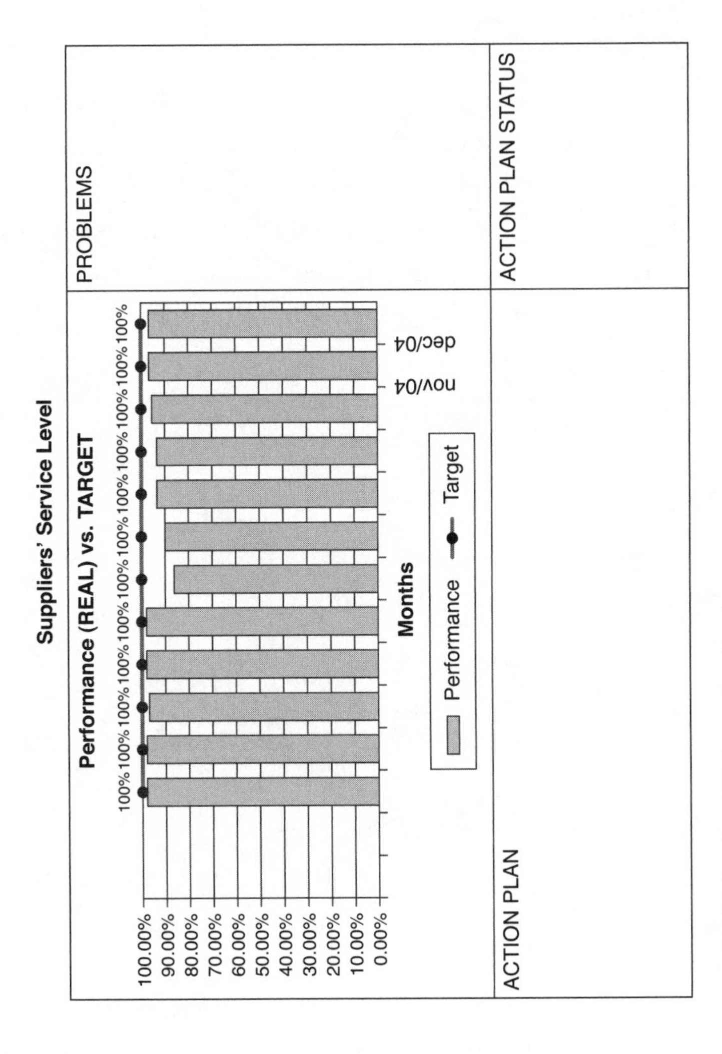

Figure 3.22 Supplier's service level

Cost avoidance: Cost avoidance is a polemical indicator. Some organizations do not accept the idea of cost avoidance. Measuring the saving which is calculated based on expected future expenses (that are managed not to happen) depends on the culture of the organization. In the case of this aerospace industry, the "cost avoidance" indicator has a specific algorithm: It measures the difference between the prices contracted versus the lowest value offered in the bid's shortlist. This example is based on data from a large multinational aerospace industry headquartered in Brazil. If the lowest-cost supplier originally offered the service for 5, and after negotiation the deal was 3, the cost avoidance is 2. Some companies question this indicator as a "high" low-price in the quotation that would distort the concept of saving.

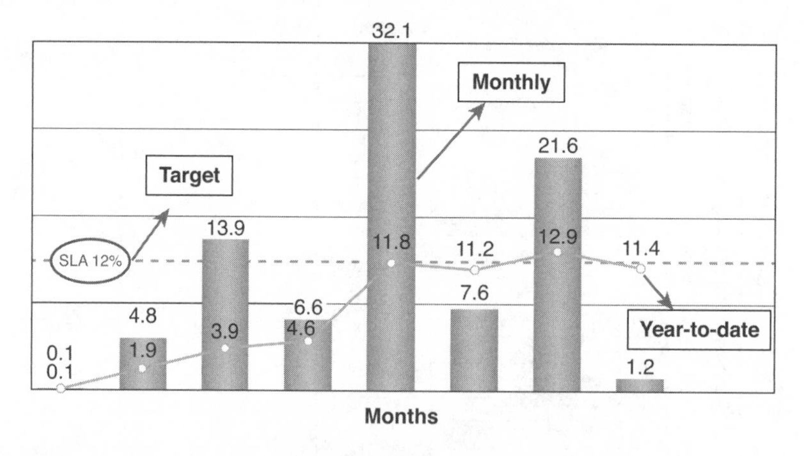

Figure 3.23 Supplier's service level

Purchase cycle: This is the time in days from purchase requisition approval to physical receiving. In the case illustrated here, the target is to reduce the purchase cycle time by 5% each quarter. This example is based on data from a large multinational aerospace industry headquartered in Brazil.

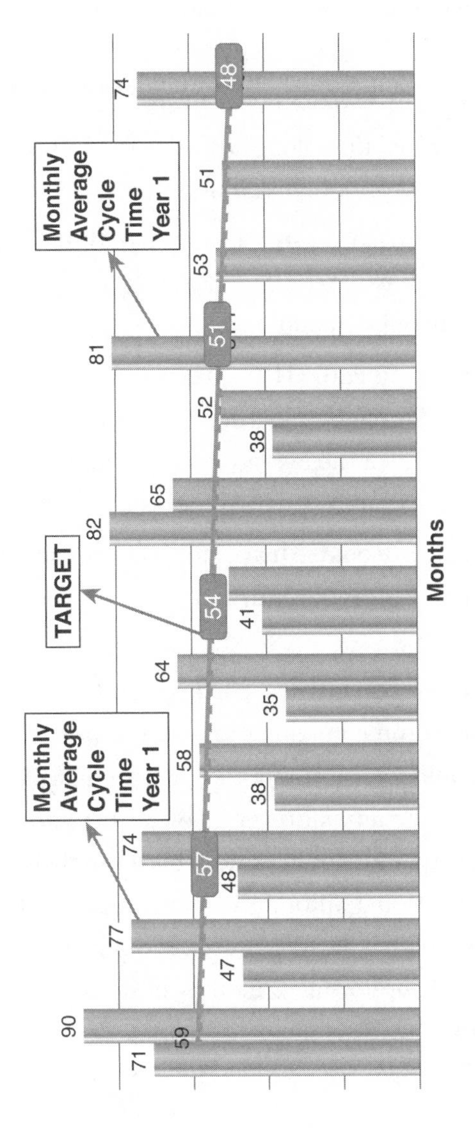

Figure 3.24 Supplier's service level

Quality of purchased parts (QPP): QPP measures both internal and customer complaints in parts per million (ppm). In this case, the customers are the automotive industries. This example is based on data from a large multinational automotive parts industry headquartered in Germany.

Inventory replenishment cycle time: This indicator measures the cycle time of production along with the time of separation from the product to the appropriate distribution center.

Gross margin return on inventory (GMROI): This is a financial metric calculated as GMROI = (Product sales value – Product cost) × Product's annual demand.

Inventory carrying rate (ICC): This is given by: [A] Total costs / Inventory costs = [Storage + Handling + Obsolescence + Failure + Administrative] / Value of the inventory = [%] and ICC = [A %] + [A %] × [(Cost of opportunity %) + (Cost of insurance %)].

Inventory carrying cost: This performance indicator is obtained by multiplying the inventory carrying rate (ICR) by the average inventory value.

Inventory turns (inventory turnover): These correspond to the number of times the company's inventory turns in the year. It is one of the most common metrics of the supply chain. A common calculation is to divide the annual revenue by average inventory cost. It can be measured in quantity or in value. This example is based on data from a large multinational automotive parts industry headquartered in Germany. Note that six turns per year is equivalent to an average inventory capable of supporting 2 months of demand. In 12 months, a 2-month inventory turns six times.

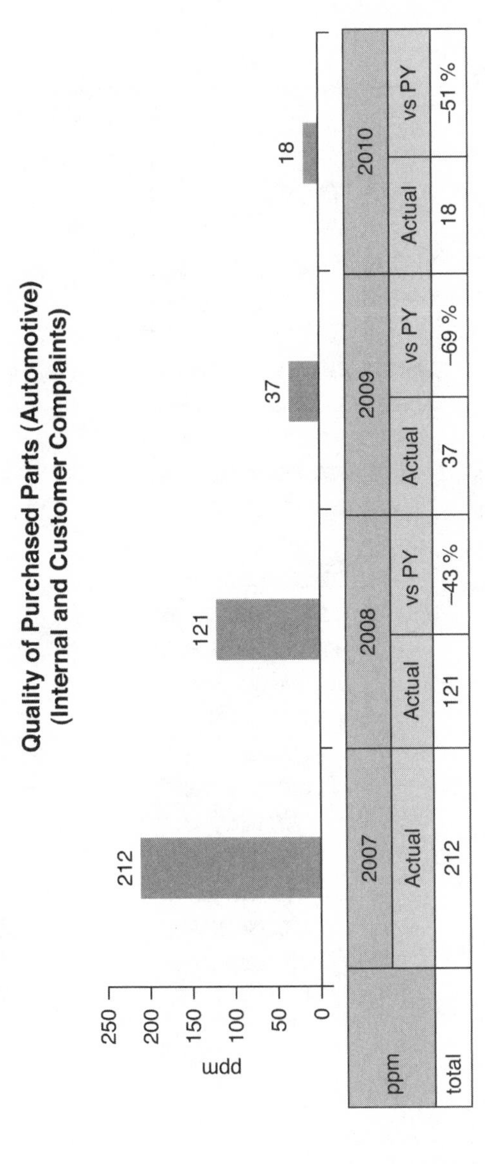

ppm	2007	2008		2009		2010	
	Actual	Actual	vs PY	Actual	vs PY	Actual	vs PY
total	212	121	–43 %	37	–69 %	18	–51 %

Figure 3.25 QPP

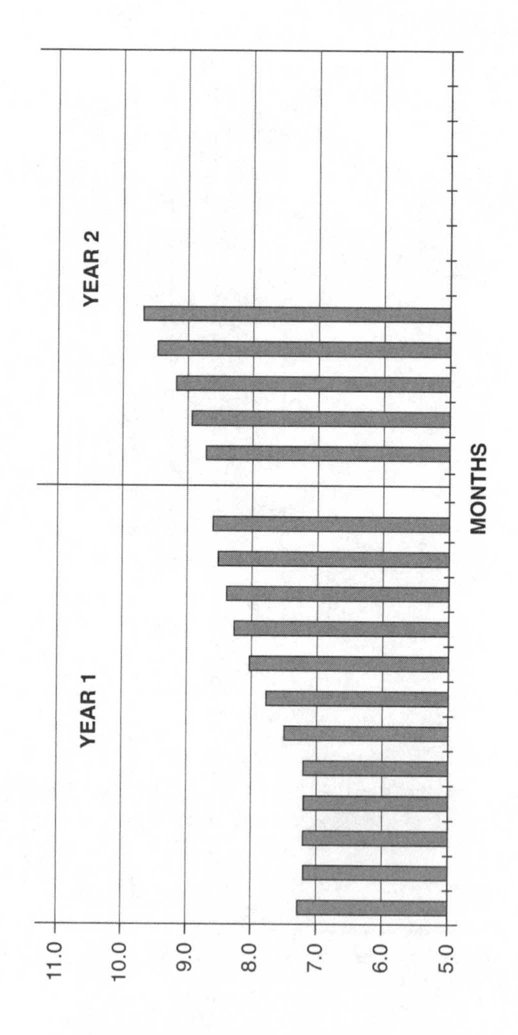

Figure 3.26 Inventory turns

Days in hand (DIH): Equivalent to inventory turns, DIH measures the inventory related to the demand (usually future demand). This example is based on data from a large multinational automotive parts industry headquartered in Italy. PA: Finished product; MA: Raw material; COMP: Parts; WIP: Work in progress; META: Target.

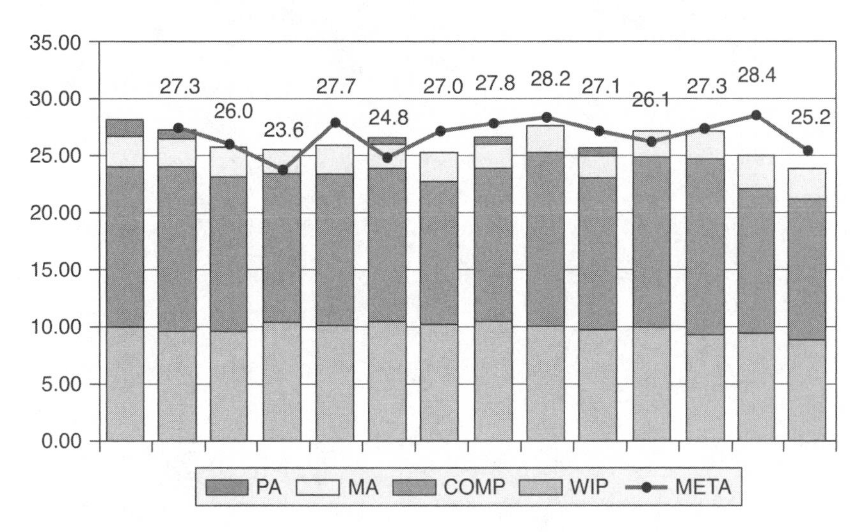

Figure 3.27 DIH

Medium-term inventory variation (MTIV): Medium-term ranges from 4 months to 2 years. Less than 4 months is considered a short time, and beyond 2 years is regarded as a long time. The control graph Figure 3.28 shows how inventory level (value, squares) reduces every month. In parallel, the line (%, dots) indicates the cumulative inventory reduction over time. This pattern follows the revision of the inventory level policy. This example is based on data from a large multinational machinery industry headquartered in the United States. Most of the inventory reduction observed in the first months occurred due to fast-moving items that held an excess of inventory; the action plan was simply to interrupt purchase orders until the inventory level was rebalanced.

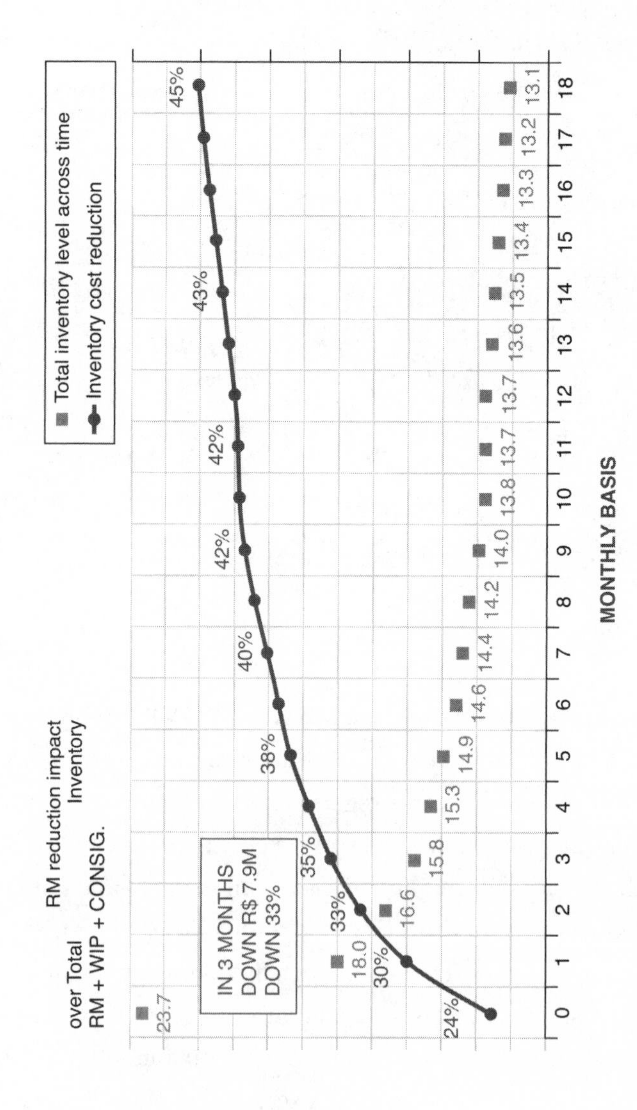

Figure 3.28 MTIV

Inventory divergence: This is obtained by the valuation of inventory adjustments carried out by month.

Inventory accuracy: This indicator relates to the receiving process and internal logistics. It aims to measure the efficiency in the control of materials and can be obtained by the ratio of the value of differences to the inventory value.

On-hold inventory: The quantity (or respective value) of items in stock but not available either for consumption (raw material) or for sales (finished product).

Stock-outs: This indicator expresses how much the stock warrants the fulfillment of requests. It can be obtained by the ratio between the number of requests without stock and the total number of orders.

Inventory variation by item: The vertical (y) axis indicates the days-in-hand (DIH) of three groups of products and the total DIH. Note that the total DIH in a given month is often smaller than the DIH of one or more categories in this same month. This happens because the total DIH is calculated by weighting the DIH of the categories. It is very likely that the volume for Industry Category is significantly higher than the others; therefore, whenever the industry's DIH is small, the total DIH is weighted downward. This example is based on data from a multinational chemical industry headquartered in France. Group of products: Painting; Industry; OEM (original equipment manufacturer).

Inventory level by origin (IBO): This control chart is similar to the previous example, but now it illustrates the DIH as per the product's origin—either imported or sourced from a local supplier. It is no surprise that imported products have a higher DIH than local items. This example is based on data from a multinational chemical industry headquartered in France.

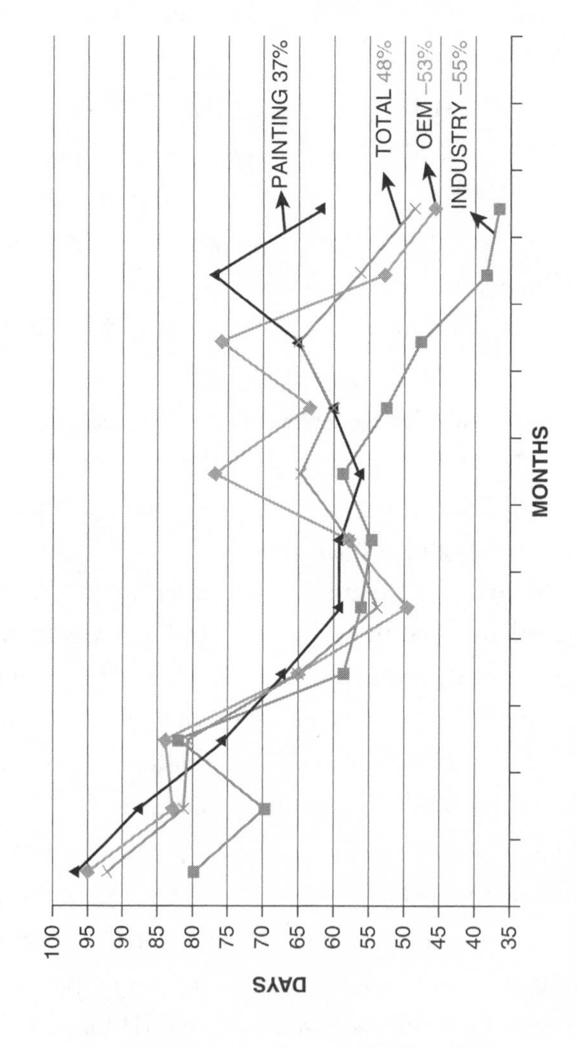

Figure 3.29 Inventory by item

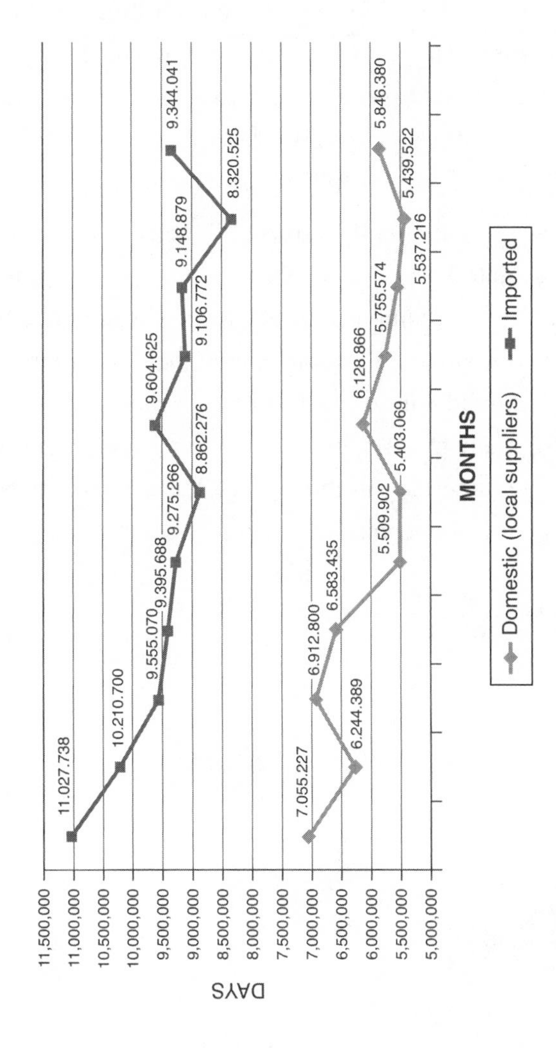

Figure 3.30 IBO

Inventory excess by origin: The value in the graph below indicates the value of inventory corresponding to the quantity of items above those planned according to the company's policy. It only indicates the excess of inventory, not the total inventory the company holds. This example is based on data from a multinational chemical industry headquartered in France. Dark bars equal inventory in excess of imported items. Light bars equal inventory in excess of locally supplied items. Axis: months versus value ($).

Working capital (WCap): Working capital is a measure of both a company's efficiency and its short-term financial health.[1] The working capital is calculated as: WCap = Current assets – Current liabilities. WCap indicates the company's underlying operational efficiency. Money that is tied up in inventory or money that customers still owe to the company cannot be used to pay off any of the company's obligations. So, if a company is not operating in the most efficient manner, it will show up as an increase in the WCap. This example is based on data from a large private hospital in Brazil.

[1] www.investopedia.com/terms/w/workingcapital.asp, accessed February 2, 2014

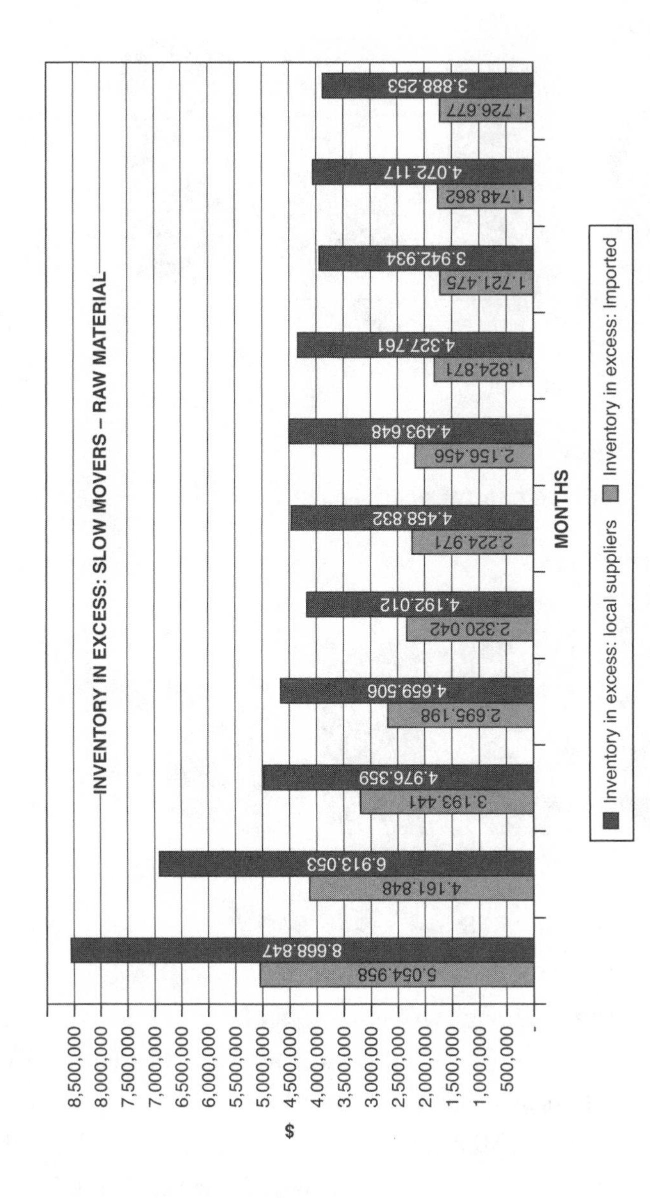

Figure 3.31 Inventory excess

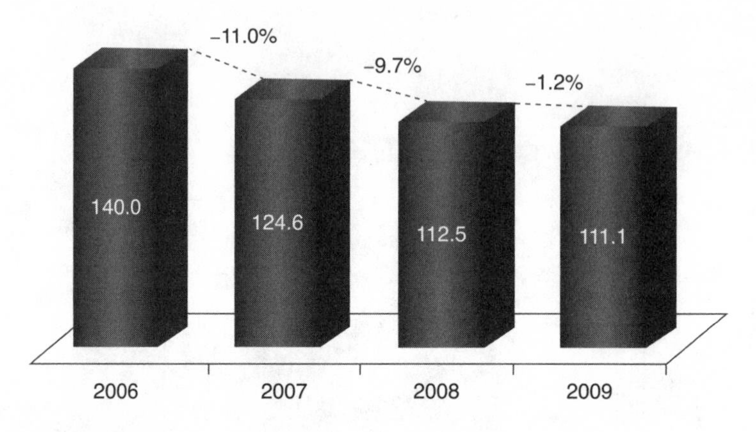

Figure 3.32 Working capital

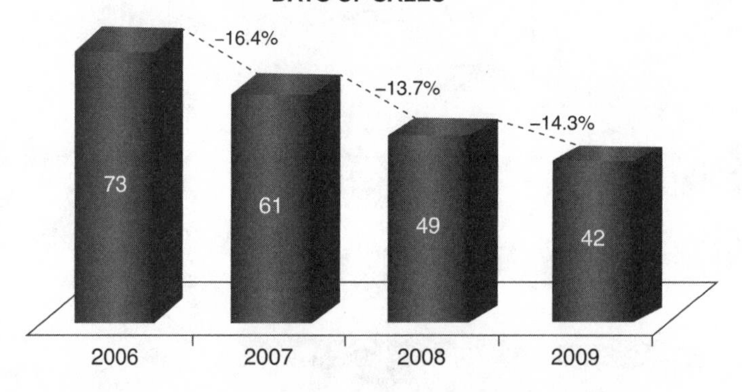

Figure 3.33 WC versus DOS

Stock aging: Although stock aging can be reported as weighted average, the best approach is to use a graphical representation with a histogram. It basically shows the period in days that a given item is in the stock location. This example is based on data from a large multinational electronics industry headquartered in the United States, in an outsourced contract manufacturing operation in Brazil. In this example, more than 400 items of raw material (not the quantity) arrived in this warehousing unit in the last 24 hours. The longer the period (moving rightward on the horizontal axis), the lower the frequency of items. It shows the profile of this stock is mostly formed by items that are quickly consumed by production.

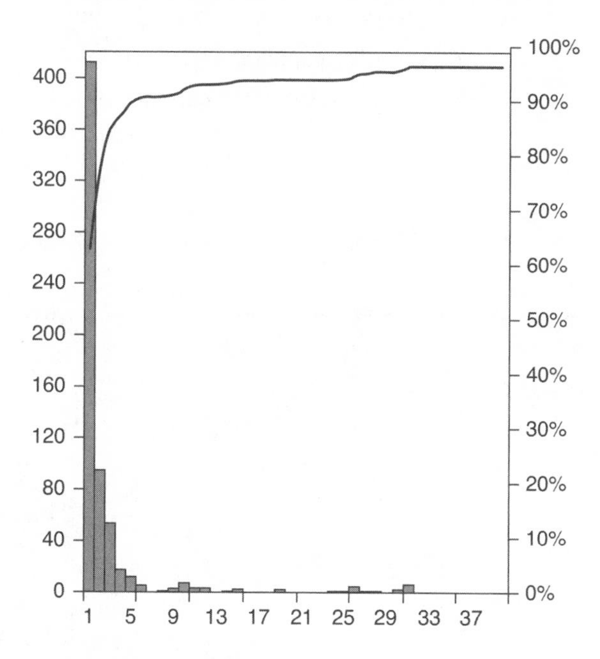

Figure 3.34 Inventory aging

Inventory by service level (INV.SL): The concept of service level requires a statistical approach according to the function name of *normal distribution service level.* It is possible to project what quantity delivers a given probability to respond to the demand. For example the quantity that delivers a service level of 98% suggests it is likely to be higher than the demand in 98% of the cases. This probability considers both demand and inventory replenishment frequency profiles. In this case, current inventory is worth $17M and is capable of delivering a 98.7% service level (SL). But the organization will only deliver this SL if the level of inventory of each item is properly balanced. This example is based on data from a medium-sized multinational metallurgic company headquartered in Italy. For this scenario, if a stock availability indicator such as LFR reports a service level much lower than 98.7%, this is probably due to an inventory imbalance at the item level. Some items have stock level increased while other a far below the required. Other factors could be an imbalanced distribution pattern and location products without a connection to regional demand areas profiles.

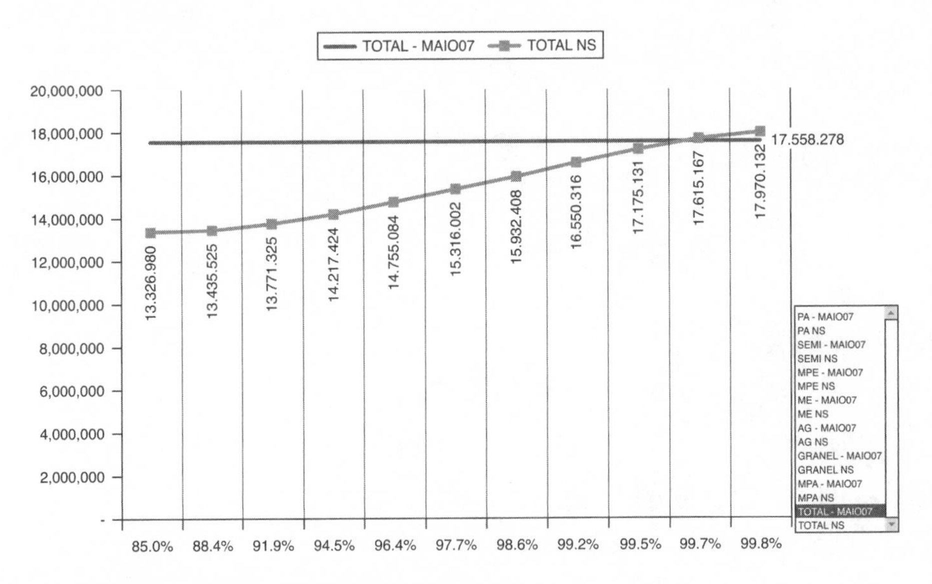

Figure 3.35 Inventory service level

Inventory by service level (INV.SL): This is continued from the previous example. Graph A: Inventory level is very low, delivers minimum service level. Graph B: Inventory level is extremely high. Graph C: Inventory level is extremely low.

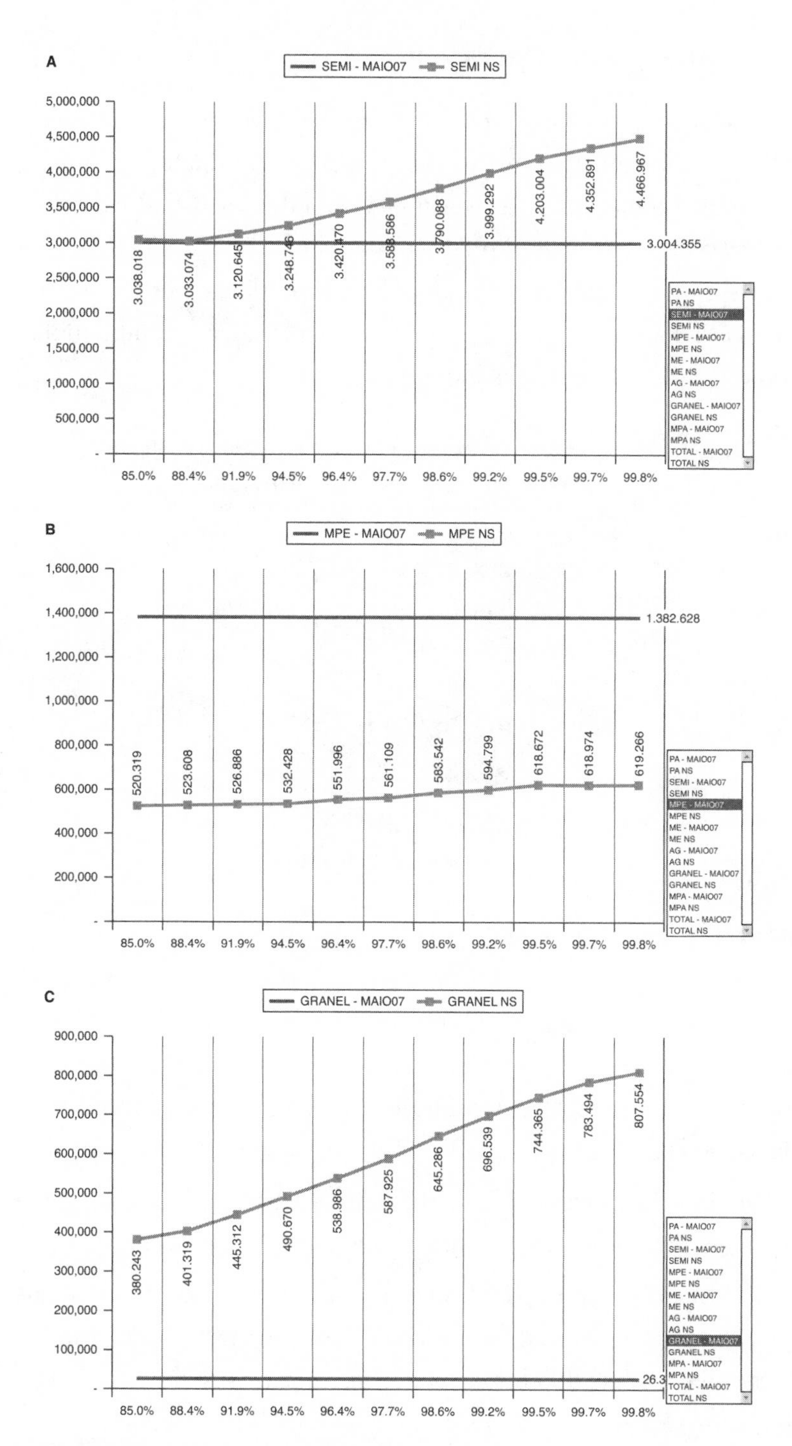

Figure 3.36 Inventory scenarios

3.2.b Synchronous Operations

Production service: This indicates the effectiveness in meeting the production. It refers to the process of production planning, which must be measured and evaluated monthly. It is defined as the ratio between the executed production and planned production. The measure can be used for items, families of products, customers and businesses. This example is based on data from a large multinational automotive parts industry headquartered in Italy.

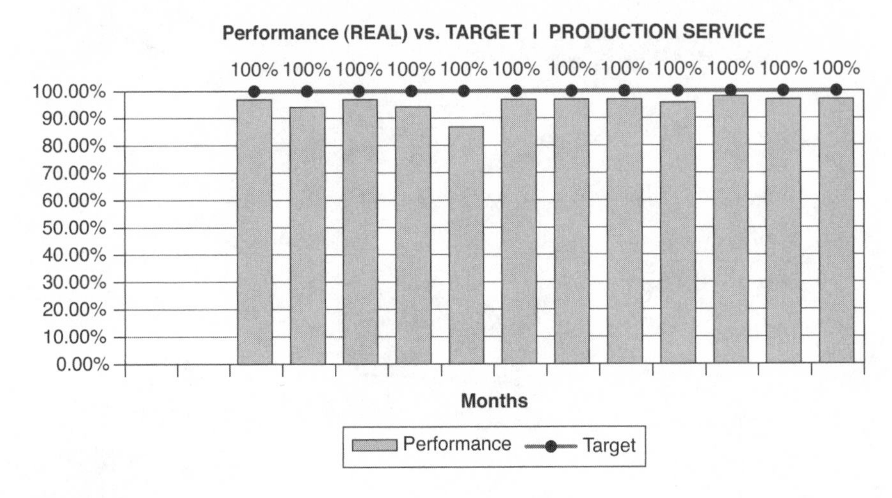

Figure 3.37 Production service

Production orders (POs): These indicate the efficiency in managing the production orders and can be calculated by the ratio between the number of POs with problems and the total number of POs issued. This indicator of efficiency is related to the area of PCP.

On-time line count: This is the total order lines shipped on time (requested date or before) versus the total number of order lines.

On-time case count: The total cases shipped on time versus total number of cases in the orders. Example: company ABC asks for 56 products totaling 200 boxes on purchase order number 1235. The manufacturer sends 140 boxes on January 3 and the remaining boxes on October 13. The requested departure date is January 3. The metric on-time case count is 70% = 140 / 200.

Delivery in full: This refers to the deliveries on time that were also perfect according to additional attributes such as having no missing cases, no damage, and so on.

Invoice matching order: Indicates the percentage of invoices without divergence. It is calculated by the ratio between the number of invoices without divergence and the total number of invoices shipped.

Inbound cost freight as a percentage of purchase: This is calculated by dividing the cost of inbound freight by the value of purchases. The result of this measurement can vary greatly.

Delivery on time (DOT): The number of orders delivered on time versus the total number of orders.

Transit time: This refers to the number of days (or hours) between the shipments of the product by the supplier until its arrival at the client.

Losses as a percentage of freight cost: These are calculated by dividing the value of the losses and failures by the total amount of freight paid in the period.

Freight invoice accuracy: This is calculated by dividing the number of invoices without errors by the total number of freight invoices during the period. Errors include incorrect price and incorrect and incomplete information.

Additional costs as a percentage of total freight cost: This is calculated by dividing the total additional costs of freight in the period. Many carriers charge extra fees for demurrage and other fees. Normally these rates represent inefficiencies in the process.

Vehicle capacity utilization: This is calculated by dividing the weight (or cubic volume or units) shipped by the total vehicle capacity.

Adherence to vehicle type: This is calculated by dividing the number of shipments sent by the ideal mode of transport by the total number of shipments in the period. To be measured, every delivery route must have defined the ideal mode of transport based on the freight cost and customer requirement.

Truck turnaround time: This is calculated by measuring the average time between the arrival of the vehicle at the establishment and the exit of the vehicle from the same establishment. It indicates the efficiency of the receiving process.

Transport tracking percentage: This is calculated by dividing the total number of shipment tracked systems by the total number of shipments during a given period.

Unplanned transport: This monitors the frequency (or corresponding value) of unplanned freights that the organization uses over time. It can be segmented into two parts: inbound freight (from suppliers) and outbound freight (to customers). This example is based on data from a large multinational automotive parts industry headquartered in Italy. Dark bars equal outbound unplanned freights. Light bars equal inbound unplanned frights. Line + Dots = Target.

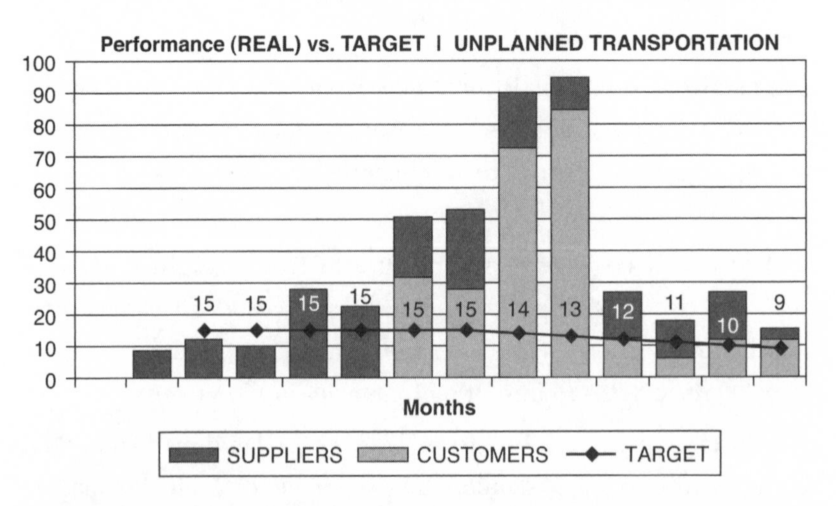

Figure 3.38 Unplanned freights

Transport cost: Transport cost can be measured and reported by various performance indicators with different calculation algorithms. The simplest way is the total expenses as a percentage of total revenue. Fortunately, other, more precise mechanisms have been developed. The following example compares the forecasted expense for August ($349) to actual results. This example is based on data from a large multinational foods industry headquartered in the United States. The distribution network split was favorable and generated $41 savings. The freight cost split also influenced the result positively. Therefore, August has reported $296, much better than the forecast.

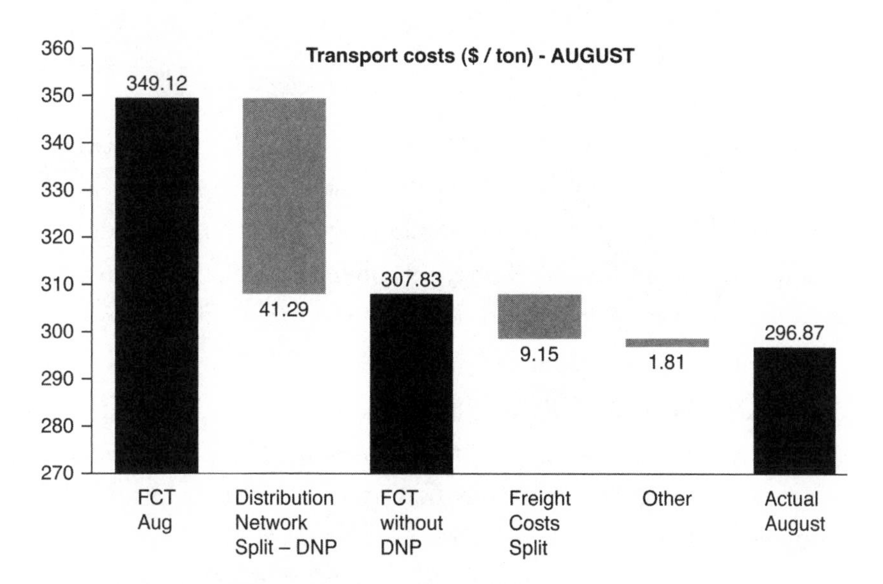

Figure 3.39 Transport cost

Load size: Like stock aging, although the load size can be reported as a weighted average, the best approach is a graphical representation with a histogram. The following graph shows that 35% of the shipments loaded more than 20 tons while only 5% were up to 3 tons. This example is based on data from a large multinational rubber industry headquartered in Italy. This approach supports better transport cost planning and also freight negotiations with carriers and logistics operators.

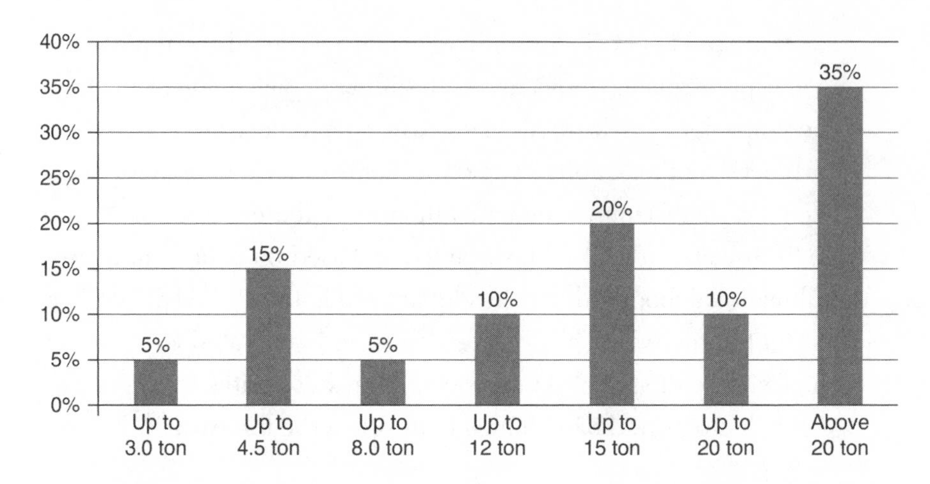

Figure 3.40 Load size

Picking on time: This indicator expresses the time between receiving an order and the end of the picking process. This example is based on data from a large multinational equipments industry headquartered in the United States. Higher line (peak nearly 100%) = Nonemergency picking on time. Lower line (peak nearly 95%) = Emergency picking on time.

Picking accuracy: Picking accuracy indicates the efficiency of the process of separation, and can be represented by the percentage of separate applications correctly performed in relation to the total number of separate materials for dispatch.

Picking productivity: This indicates the efficiency of the process of separation of material per hour which can be obtained by counting the number of separate boxes per man-hour. The example is based on data from a large multinational equipments industry headquartered in the United States. Light bars equal the number of working hours (WH). The dotted line represents the goal. The continuous line represents the lines/hour of productivity. The leftmost bar is the Average WH = 15,210 / Average productivity = 8.8.

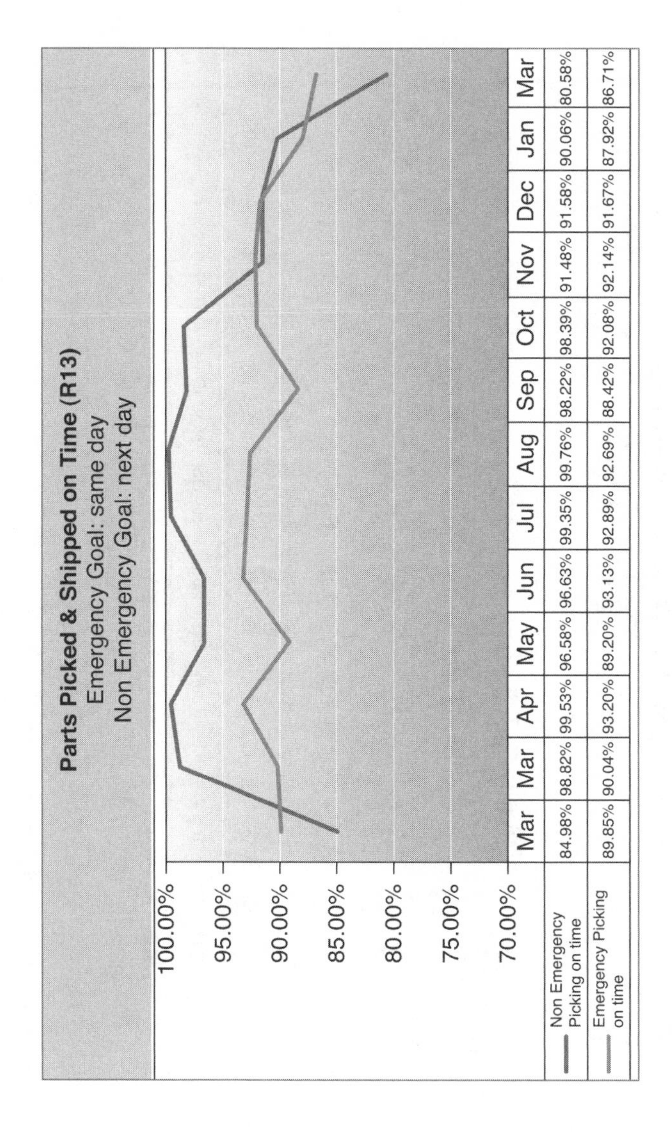

Figure 3.41 Picking on time

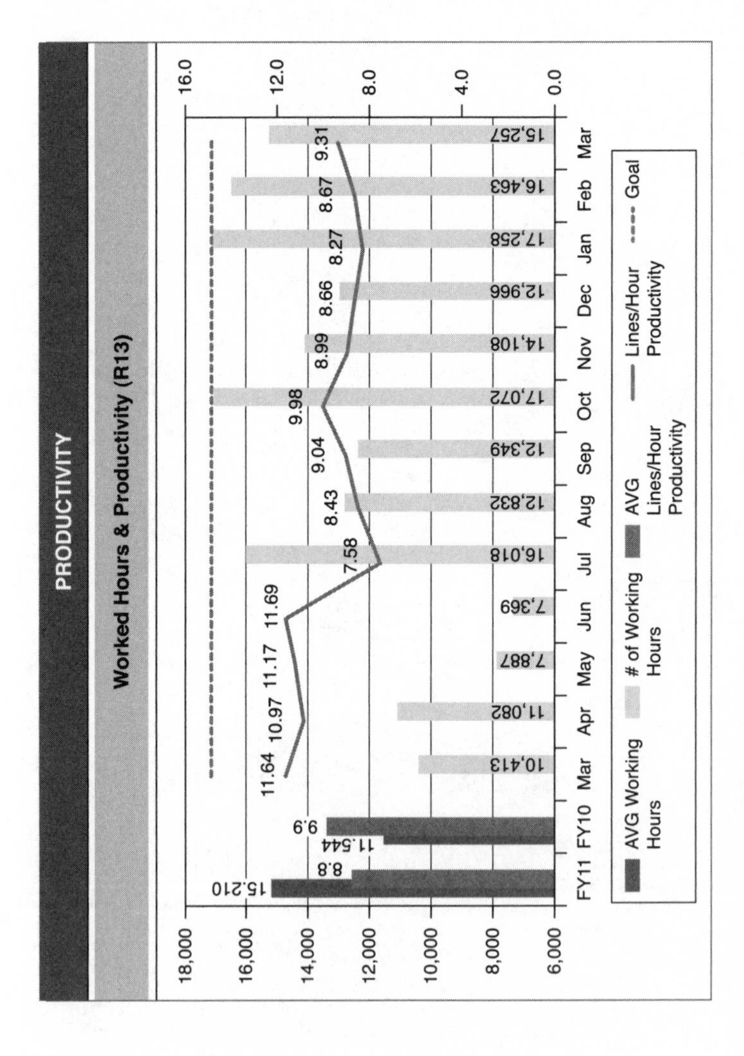

Figure 3.42 Picking productivity

Loading time: Loading time indicates the duration of the activity of loading the vehicle for shipment. It can be calculated as the difference between the end of charging time and the time of commencement of loading.

Receiving without divergence: This metrics indicates the percentage of material that do not show physical divergence versus the total number of invoices.

Warehousing productivity: This indicates the storage capacity of a worker and can be obtained by counting the number of pallets stored by one man in 1 hour.

Team productivity: Team productivity indicates the productivity of a worker in an operation (receiving, picking, shipping, and so on). This example is based on data from a large multinational chemical industry headquartered in France. The graph indicates the number of cases shipped from a warehouse per operator per hour over the months.

Shipping productivity: Shipping productivity aims to indicate the efficiency of the resource management in the dispatch process. This can be obtained by the ratio of the number of items shipped and the number of hours worked, thus obtaining an indicator of efficiency of items per hour.

Vehicle check-in time (VCT): VCT is defined by the period each vehicle waits, from arrival at a warehousing facility to the moment the unloading process starts. The following histogram indicates that although the average VCT is slightly above four hours, several vehicles waited even more than 10 hours. This example is based on data from a large multinational household retailer headquartered in France.

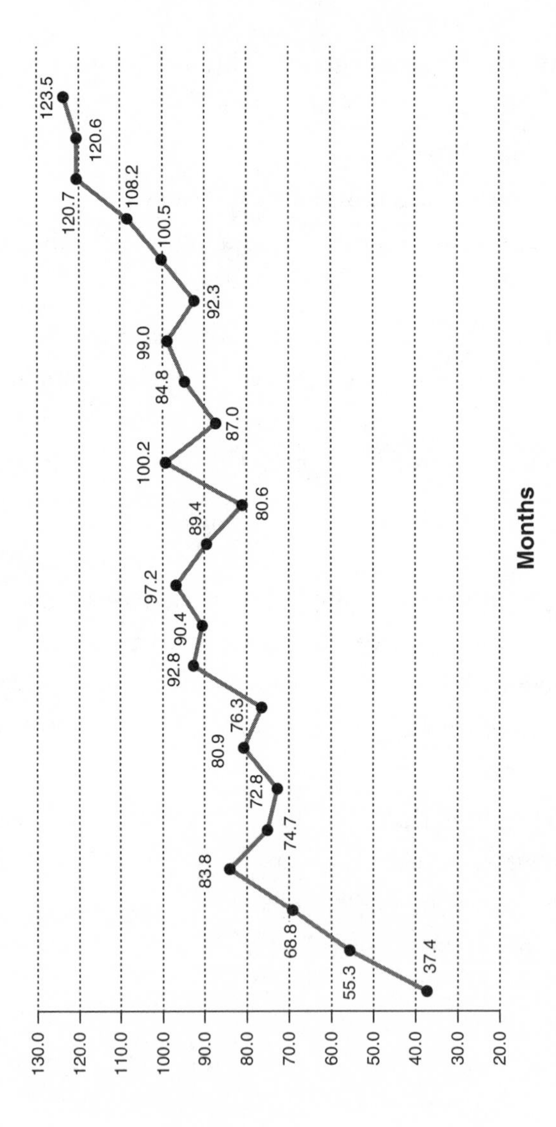

Figure 3.43 Team productivity

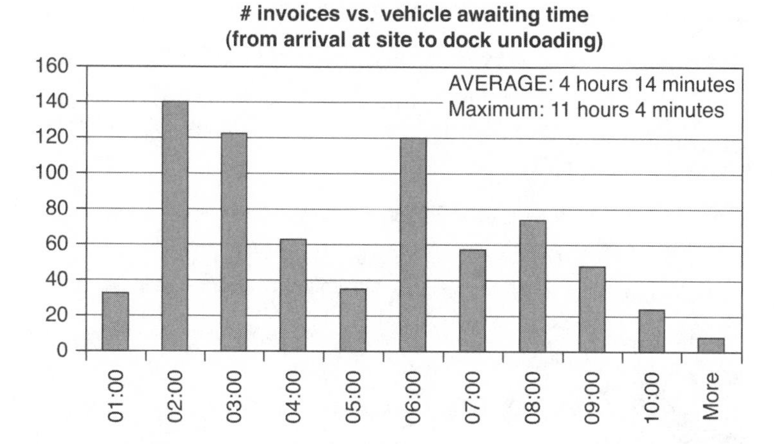

Figure 3.44 VCT

Vehicle unloading time (VUT): VUT is defined by the unloading period. The following histogram indicates the unloading process was completed in less than 1 hour for most of the vehicles. This example is based on data from a large multinational household retailer headquartered in France.

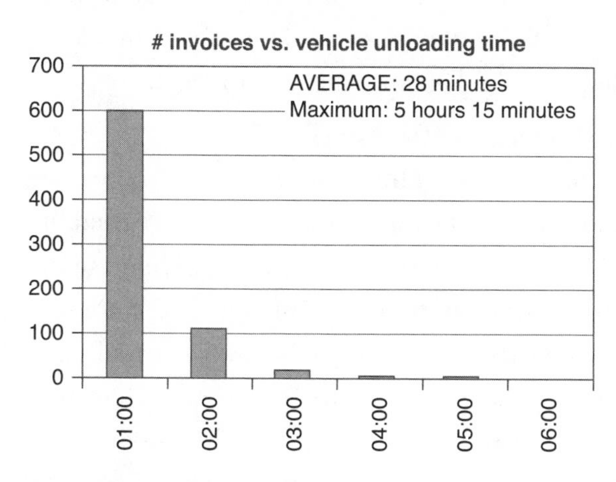

Figure 3.45 VUT

Total vehicle staying time (TVST): TVST indicates the period from vehicle arrival until it leaves the warehousing facility. This example is based on data from a large multinational household retailer headquartered in France.

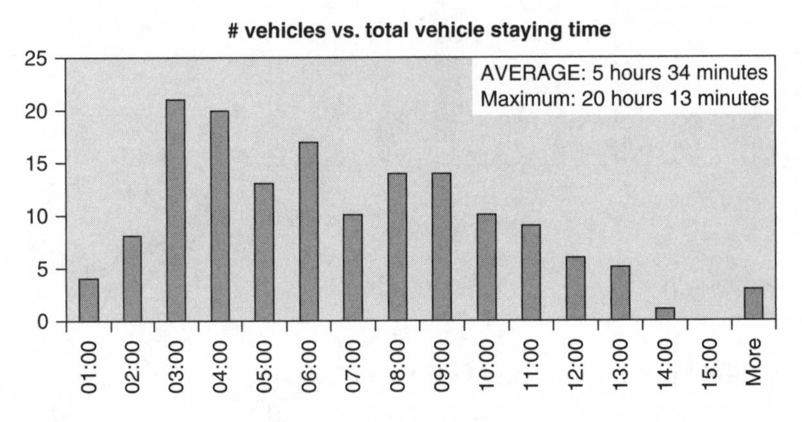

Figure 3.46 TVST

Total receiving time (TRT): TRT indicates the period from vehicle arrival at the warehouse until it finished the unloading process. RT = VCT + VUT. This example is based on data from a large multinational household retailer headquartered in France.

Inventory record accuracy (IRA): This indicator represents the level of reliability there can be in the inventory records. It can be obtained from the ratio between the number of correct scores and the number of counts carried out. A count can take into consideration the type of product, quality, batch, position, and amount. There are several algorithms for this indicator. This example is based on data from a regional automotive parts retailer headquartered in Brazil. The target varies from 90% (January) to 96.5% in December. The report shows evolution but it has been below target since July.

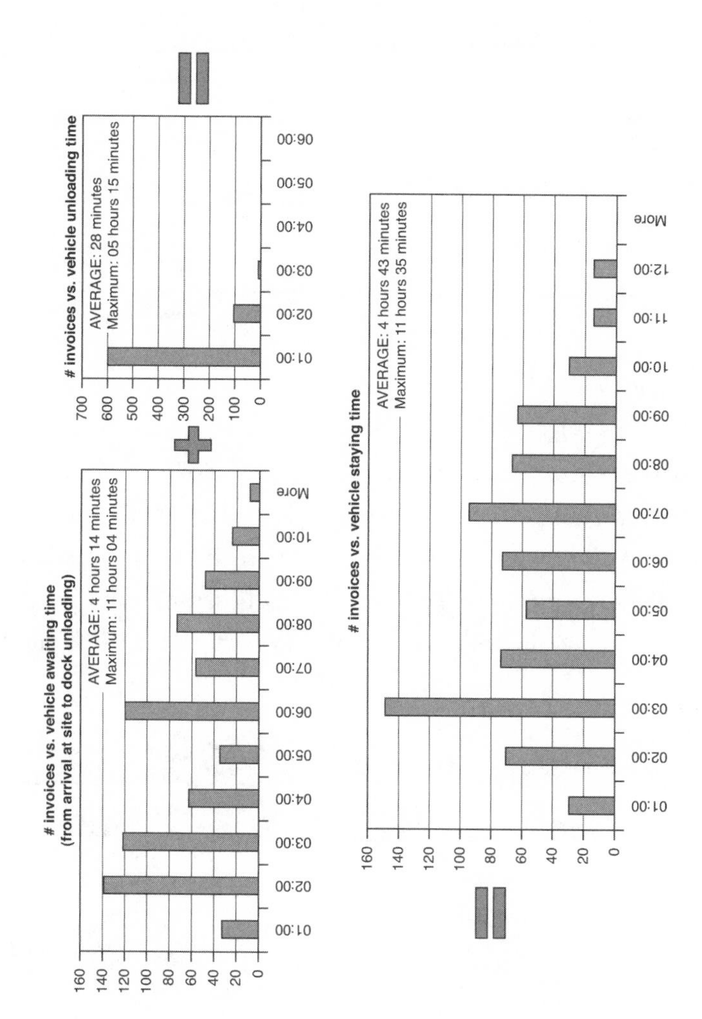

Figure 3.47 TRT

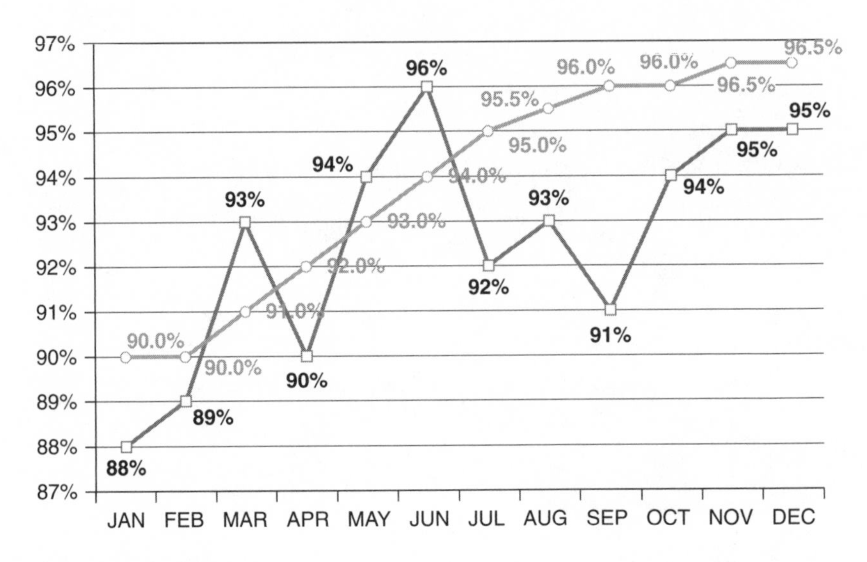

Figure 3.48 IRA

Shipping errors: These are calculated by dividing the number of errors (that is, cases shipped to the wrong customer) by the total cases shipped. As happens in other metrics, there are many variations for the calculation algorithm. This example is based on data from a large multinational equipments industry headquartered in the United States.

QUALITY

Shipping Errors/1000 (R13)

	FY11	FY10	Mar	Apr	May	Jun	Jul	Aug	Sep	Oct	Nov	Dec	Jan	Feb	Mar
AVG Report Date	7.6	6.0													
AVG Shipped Date	6.7	5.4													
Report Date			6.7	4.3	4.7	4.1	8.0	5.1	6.0	5.5	8.2	9.7	6.4	5.9	7.6
Shipped Date			5.1	3.2	5.0	5.9	5.8	5.8	5.2	3.5	8.3	8.0	6.5	6.5	4.1
Goal			2.7	2.7	2.7	2.7	2.7	2.7	2.7	2.7	2.7	2.7	2.7	2.7	2.7

Figure 3.49 Shipping errors

Warehousing occupation: This expresses the degree of occupation of the warehouse. It is indicated by the ratio of the number of positions held in the warehouse and by the total number of positions. There are several variations in the calculation algorithm. This example is based on data from a large multinational chemical industry headquartered in Brazil.

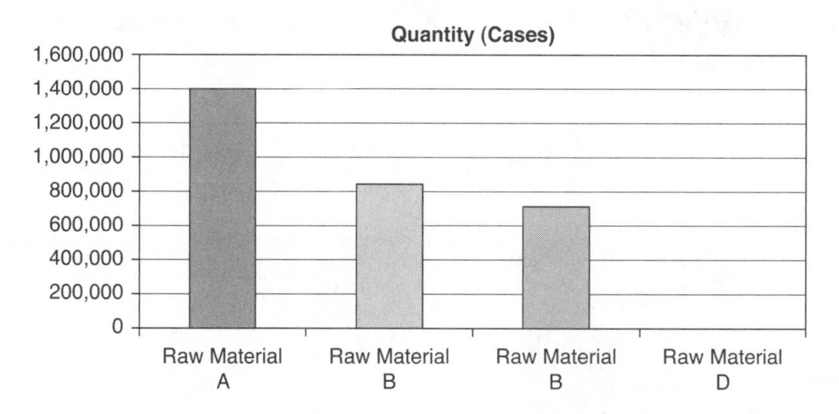

Figure 3.50 Warehousing occupation

Receiving seasonality: This indicates the volume of vehicle arrivals over time.

Dock-to-stock cycle time: This indicates the time of the storage operation from the vehicle dock until the stock is unloaded. It can be calculated as the difference between the time of unloading the vehicle and the time at which the product is stored.

Forecasting warehousing occupation: This expresses the degree of occupation of the warehouse over time. It considers the expected business volume variation. This example is based on data from a large multinational chemical industry headquartered in the United States.

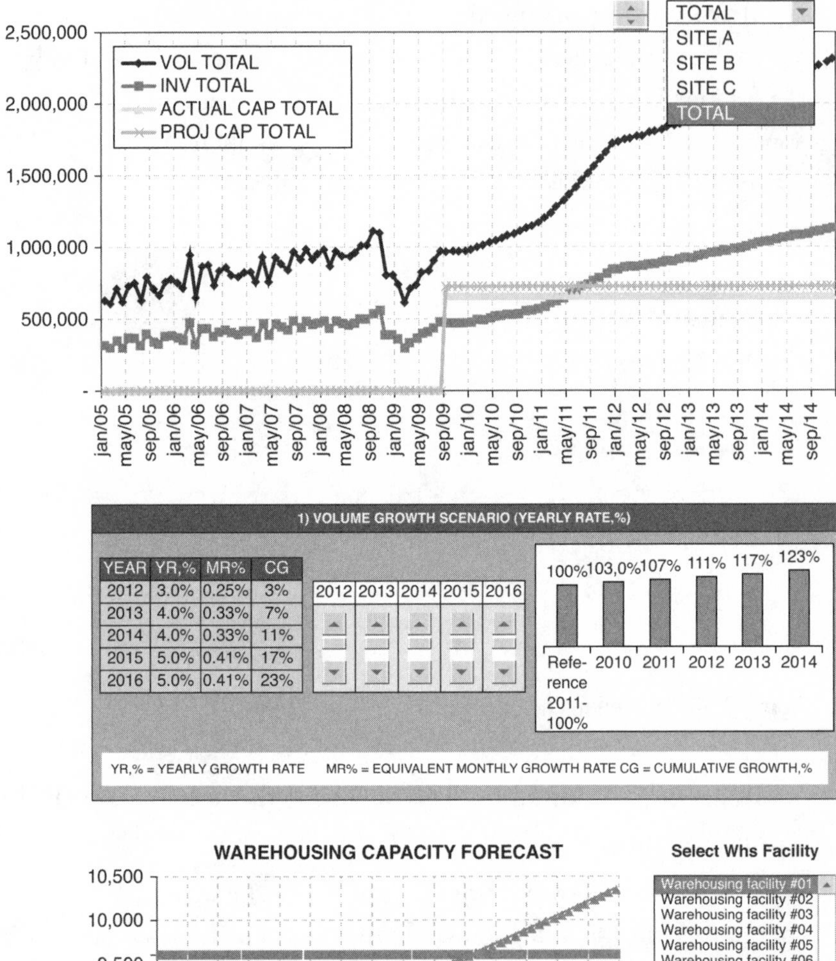

YEAR	YR,%	MR%	CG
2012	3.0%	0.25%	3%
2013	4.0%	0.33%	7%
2014	4.0%	0.33%	11%
2015	5.0%	0.41%	17%
2016	5.0%	0.41%	23%

Figure 3.51 Forecasting capacity

- Higher inclined line: Volume forecast.
- Second inclined line: Projected inventory level.
- Flat lines: Actual and future warehousing capacity.
- Simulation panel: Projects sales volume for the following years.
- Based on sales volume forecast, it is possible to project the inventory level over the time.
- Lower flat line: Current warehousing capacity.
- Higher flat line: Future warehousing capacity.
- The difference is capable of absorbing any inventory increase for nearly 12 months.

Receiving schedule adherence (RAA): RAA expresses the degree of compliance with the programming of the receipts. It can be expressed by the ratio of the number of receipts within the delivery window, and total receipts. This example is based on data from a large multinational household retailer headquartered in France.

Lower-left graph: late arrivals. Lower-right graph: anticipated arrivals. Pie chart: indicates 39% of vehicle arrivals occur after the scheduled receipt window.

Cost per order: This indicates the operating cost of the warehouse by order processed.

Warehouse handling cost: This indicates the percentage of the cost of materials handling in the warehouse as against gross sales. Some other calculation algorithms are more precise. Ideally, the use of activity-based costing is recommended.

Obsolescence cost: This is indicated by the cost of obsolete materials discarded monthly.

Returns: Returns refer to the volume returned. They can be measured against total revenue, or as a percentage of shipped volume or using many other mechanisms. This example is based on data from a large multinational foods industry headquartered in Switzerland.

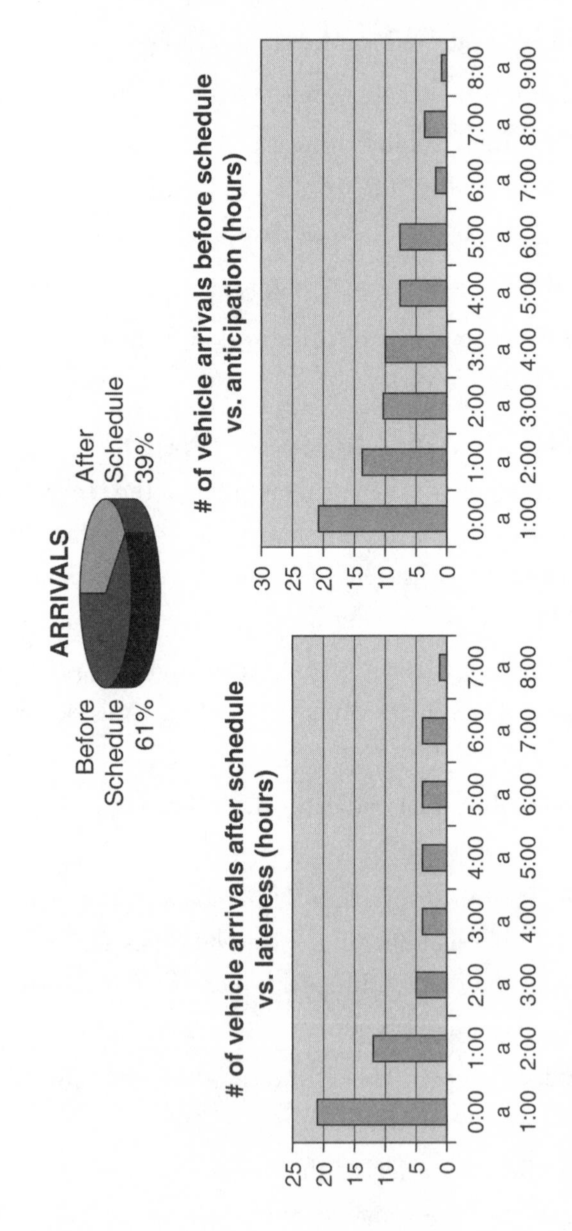

Figure 3.52 RAA

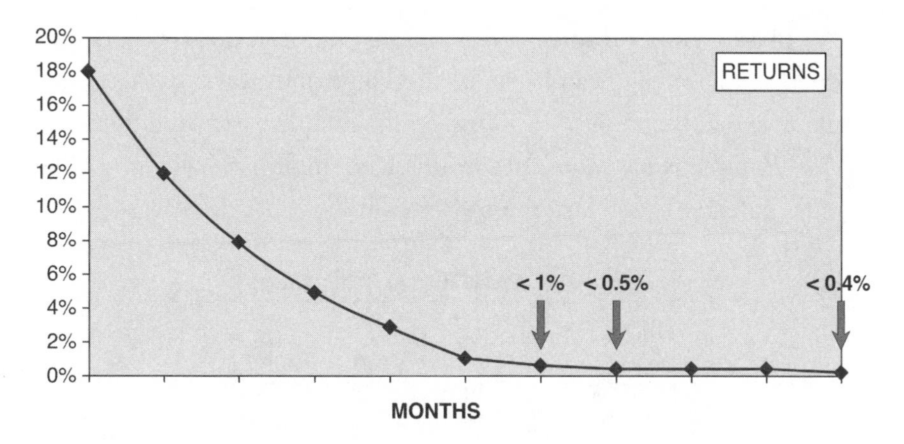

Figure 3.53 Returns

Returns by segments: Similar to the previous performance indicator, this only differs in that it segments the overall index. In this example, the segmentation occurs by salesperson but there are other possibilities such as region, distribution channel, product, shipping warehouse, etc. This example is based on data from a large multinational foods industry headquartered in Switzerland.

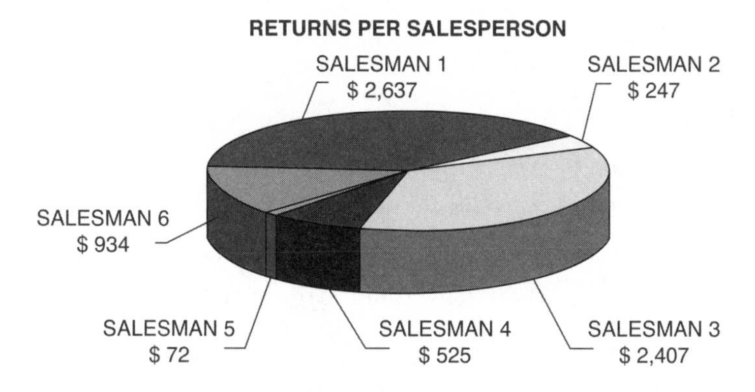

Figure 3.54 Returns, segmented

Production cycle time: This is measured from the "firm planned order" until the final production is confirmed. Generally, it takes into consideration the amount originally planned versus the quantity actually produced.

Production volume: A very simple indicator, this registers production volume and can be segmented in various ways. In this example, it shows the monthly average production that occurred per shift. The example is based on data from a large multinational automotive parts industry headquartered in Germany.

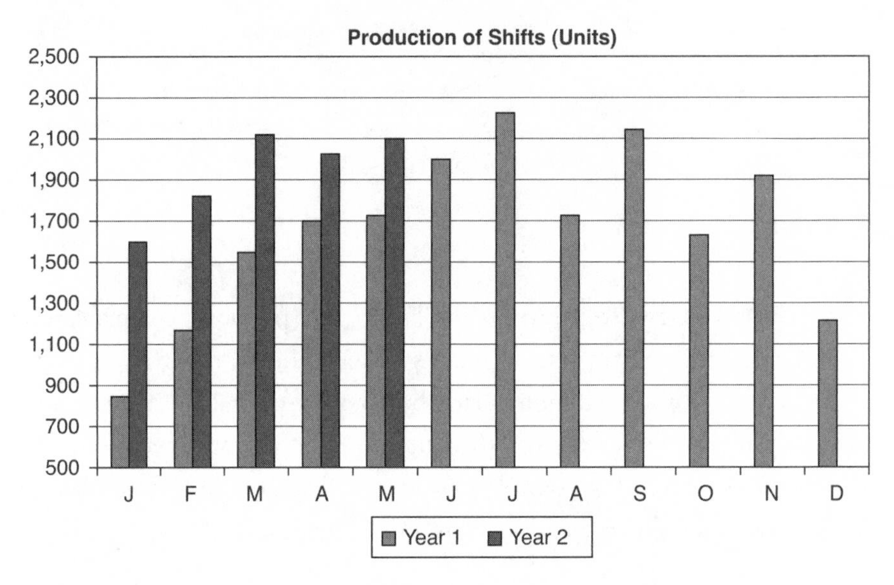

Figure 3.55 QPP

Freight cost per shipped unit (FCPSU): This is calculated by dividing the total cost of freight by the number of units shipped in the period. This metric is very useful in business where the units of measurement are standard. It can also be calculated by mode of transport. This example is based on data from a large national newspaper industry headquartered in Brazil.

Outbound cost freight as percentage of net sales: This is calculated by dividing the cost of outbound shipping by net sales. Most accounting systems can separate freight arrival and leaving. The outcome varies depending on the mix of sales, but is an excellent indicator of the financial performance of the transport.

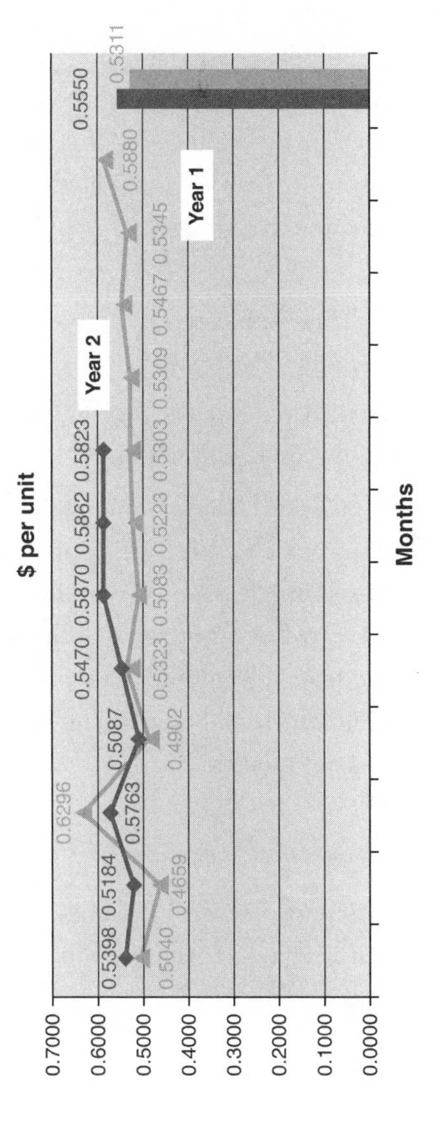

Figure 3.56 FCPSU

Outbound freight: Outbound freight can be obtained by the ratio between the value of the freight and the value of the goods transported.

Export freight cost: This can be obtained by the ratio between the value of the freight and the value of the exported goods.

Shipping seasonality: This illustrates the volume shipped over time and can be segregated in different ways. This particular example reposts the shipping seasonality per product. The example is based on data from a large multinational beverage industry headquartered in the United States.

Export lead time: This is obtained by the average of the total time spent in days, reviewed by destination.

Import freight cost: This cost can be obtained by the ratio between the value of the freight and the value of the imported goods.

Distribution service level (DSL): This indicator reports the level of service offered, for which there are various calculation algorithms available. In this example, the organization compares the performance of two 3PL providers that serve different regions. The example is based on data from a large multinational payment service industry headquartered in the United States. The bars represent the consolidated service level per quarter.

Case: Monitoring the Transport Operation

This section presents performance indicators for fleet management based on data from a regional furniture industry with a fleet of 110 trucks headquartered in Brazil.

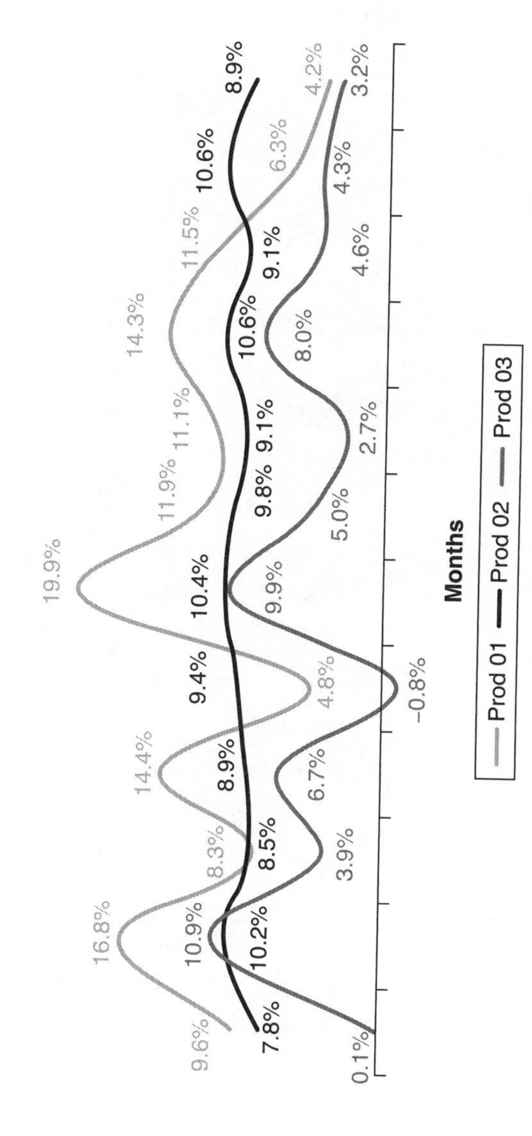

Figure 3.57 Shipping seasonality

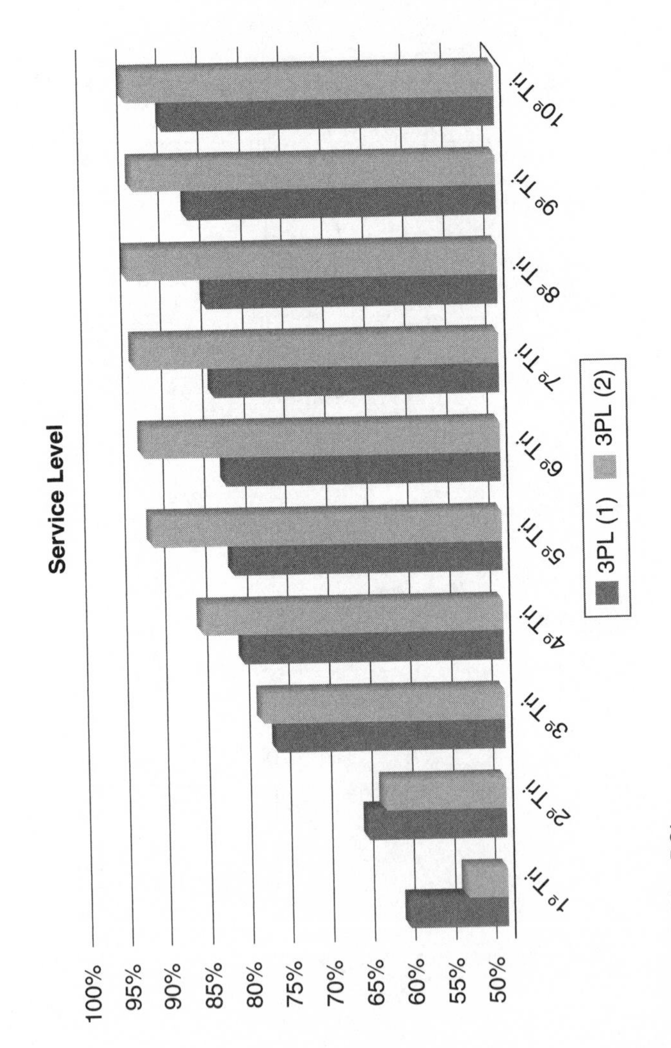

Figure 3.58 DSL

Trips per driver: The chart follows a pattern within the annual average (3.32), without great anomalies, except for January and February where the number of loads and travel are naturally smaller as high season period is in the last quarter of the year.

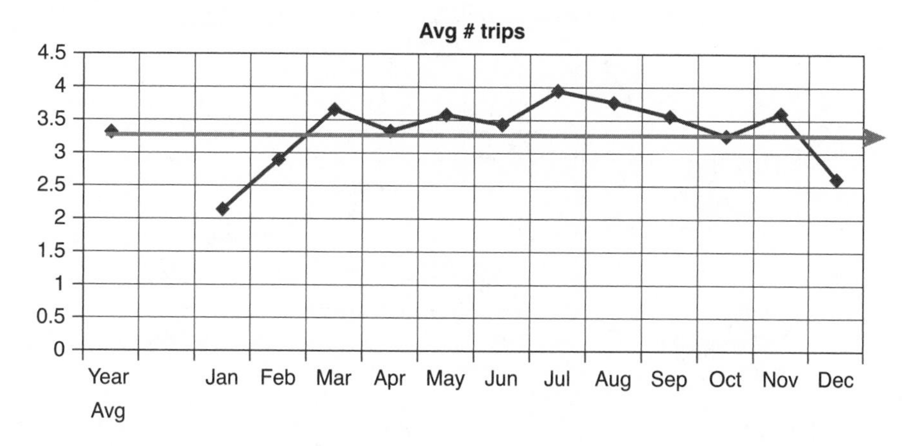

Figure 3.59 Trips per driver

Deliveries per driver: The chart follows a pattern within the annual average (51.22), without great anomalies, except for January and February where, again, the number of loads and travel are naturally smaller.

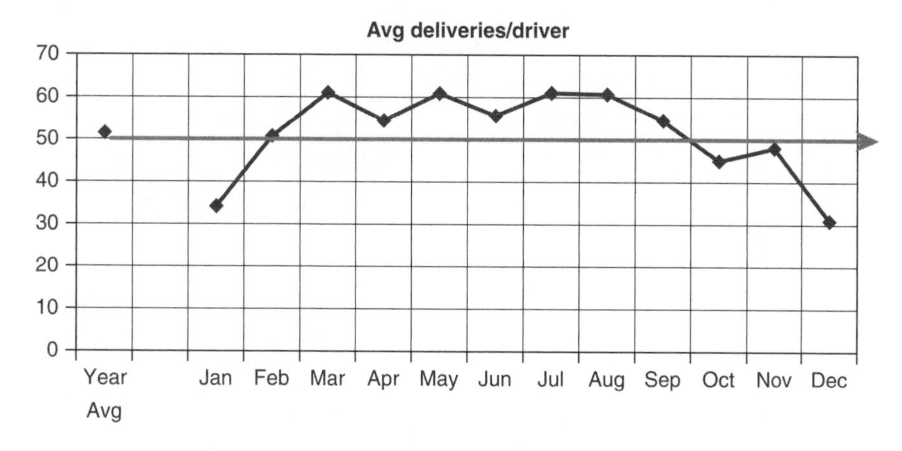

Figure 3.60 Deliveries per driver

Mechanical failure index (MFI): Starting in June, there has been a change in the pattern, leading to a reduction in the number of broken parts in relation to the number of pieces transported.

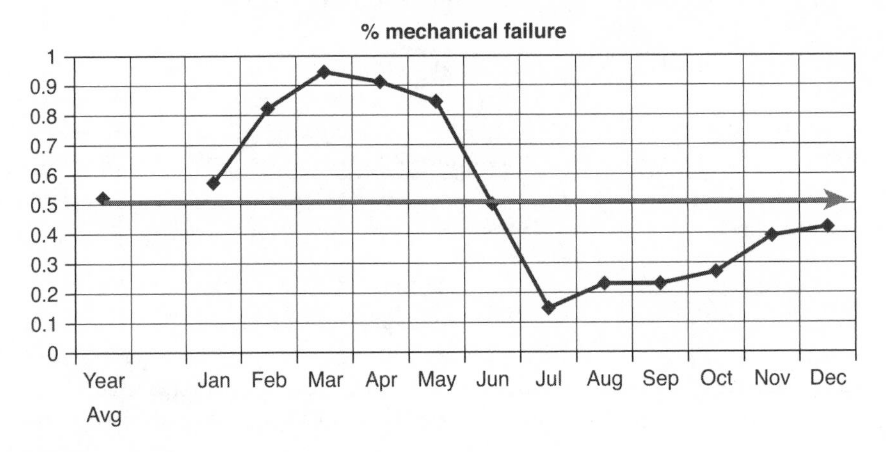

Figure 3.61 MFO

KM/liter per driver: There has been a change of pattern from September, leading to a lower average. That arose from maladministration in the choice of trucks, where trucks that consumed more fuel made trips that could have been made by more economical vehicles.

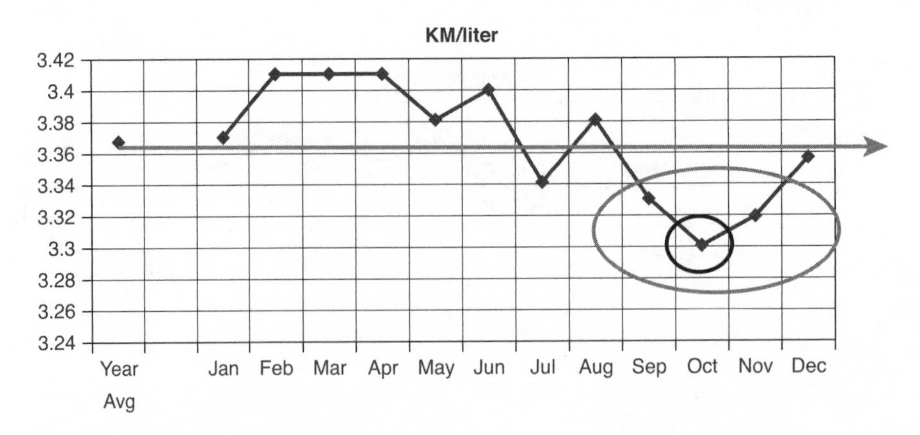

Figure 3.62 Km/L per driver

Number of days on the road per driver (DOR): This chart follows a pattern within the annual average (18.8), without great anomalies, except for January and February where the number of loads and travel are naturally smaller.

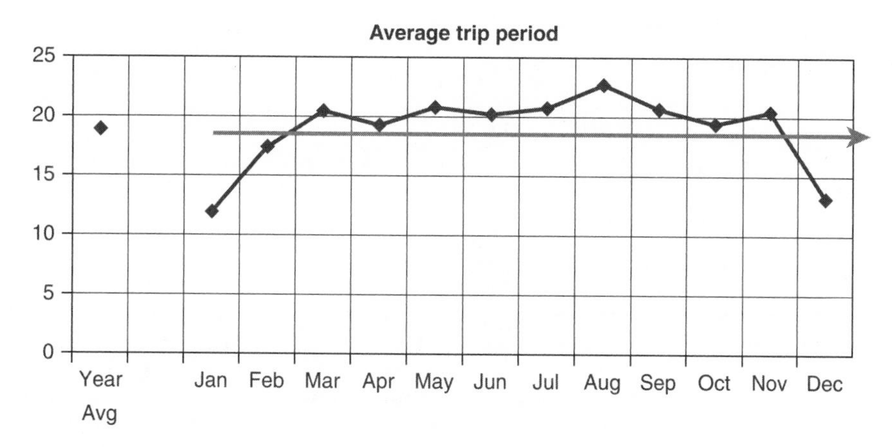

Figure 3.63 DOR

Speed excess: At the beginning of the year, the excess speeds recorded by the onboard computer of the trucks are below average; from May those numbers are very close to the average, except for 2 months. The average excess per trip is 38.

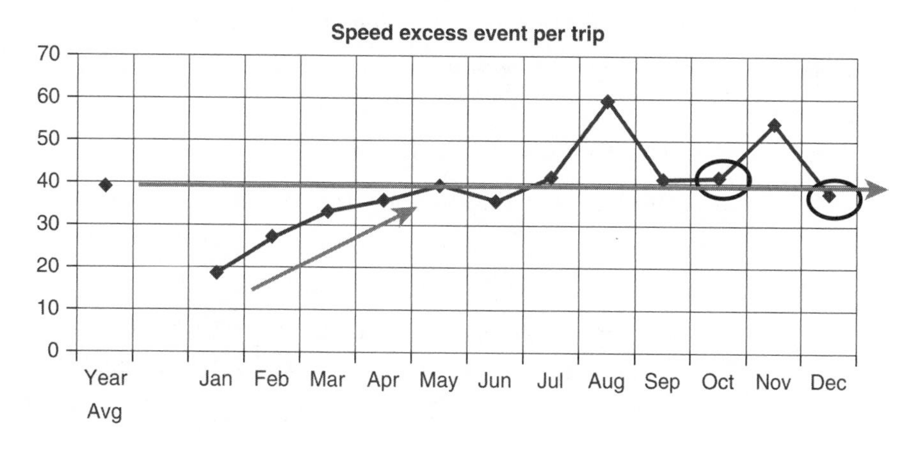

Figure 3.64 Speed excess

Speed excess per driver: The chart indicates the number of times during a given month each driver exceeded the speed limit. The drivers in the left are the most problematic.

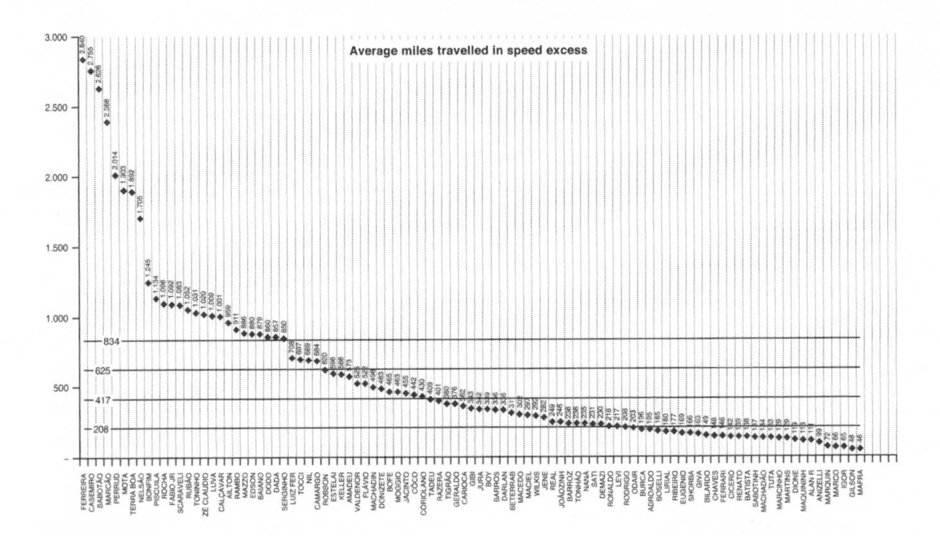

Figure 3.65 SE per drive

3.2.c Tactic Governance

Service level and inventory: This indicator compares the service and inventory levels. These patterns will not necessarily behave equally. If the organization is improving the quality of the inventory, an inventory level reduction may occur while the service level increases. This example is based on data from a large multinational chemical industry headquartered in Denmark.

Region A: Reduction of stock level from 45 days in hand (Sep 07) to 32 (Sep 08) while service levels improve from 94% (Oct 07) to nearly 100% (Sep 08). The service level difference from Sep 07 to Oct 07 is due to a change in the calculation algorithm. Region B: The service level continued to oscillate from 98% to 100% while the inventory level dropped from 580 days in hand to 110.

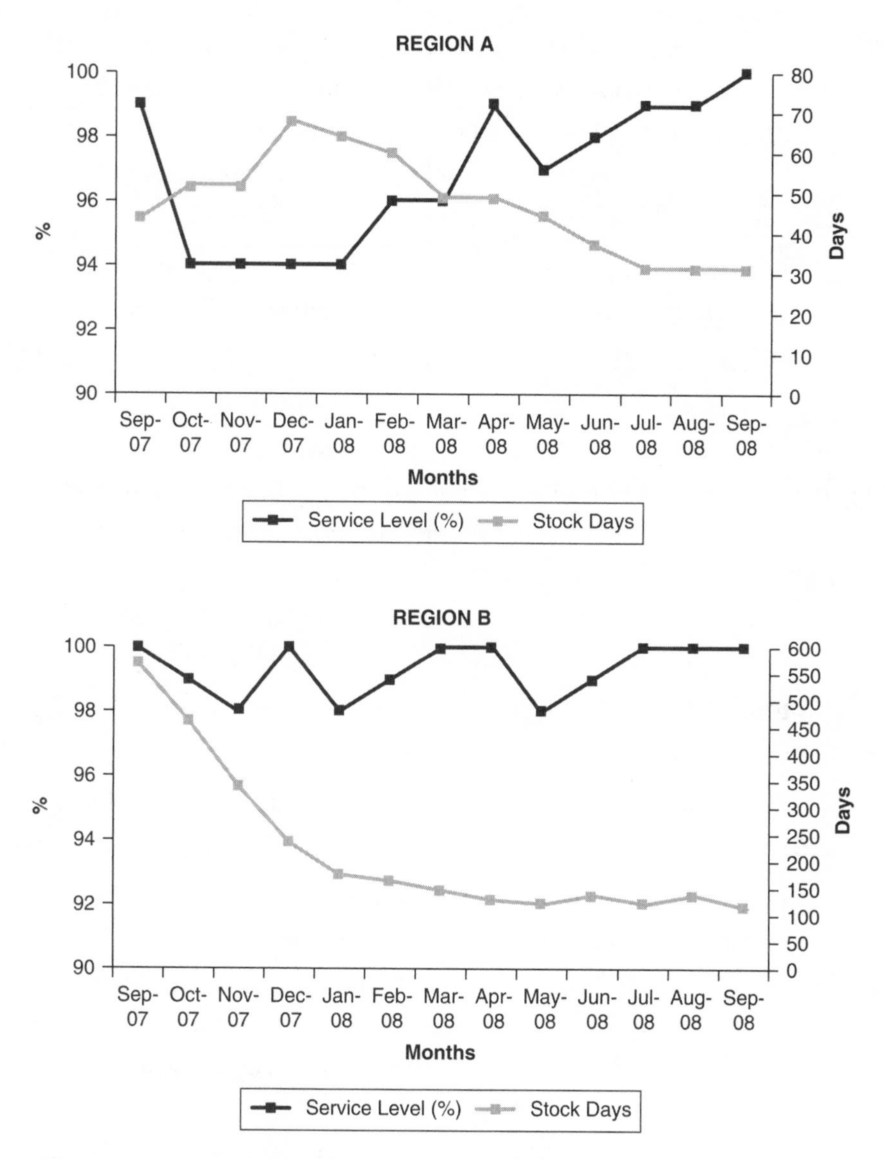

Figure 3.66 DIH versus SL

Customer order promised cycle time: This is the time promised from the creation of the order to its delivery.

Customer order actual cycle time: This is the actual time from the creation of the order to its delivery.

Line fill rate (LFR): The LFR is the number of order lines available in stock versus the total number of order lines requested. Example: A company receives orders for one unit of ten items (total of ten order lines). The manufacturer only ships seven items (seven lines). The fill rate is therefore 70%.

SKU fill rate: This is considered to be the number of SKUs (stock keeping units) shipped versus the number of SKUs ordered.

Order size: This indicator tracks the order size, usually against a target defined by a maximum expected quantity. Alternatively, the organization may wish to track the order size variation; this may be done by using the order size standard deviation, as in the following example. This example is based on data from a large multinational FMCG industry headquartered in the United States and serving a key account retailer in the United Kingdom. The target is to keep order size variation, given by the standard deviation, below 60%.

Case fill rate: This fill rate is the number of cases shipped versus the number of cases ordered.

Value fill rate: The indicator is the same as in the case fill rate except for considering the value of each line of the order.

DIH, OTIF, and inventory value: This example illustrates the combined use of different performance indicators over a 4-year horizon. The DIH (as previously explained) was 51% above target in year 1 and consistently reduced until it reached target in year 4. This example is based on data from a medium-sized home appliances industry headquartered in Brazil.

Customers' claims: These indicate the dissatisfaction of customers through the number of complaints received. In this example, complaints are compared versus total units delivered. This example is based on data from a large national newspaper industry headquartered in Brazil.

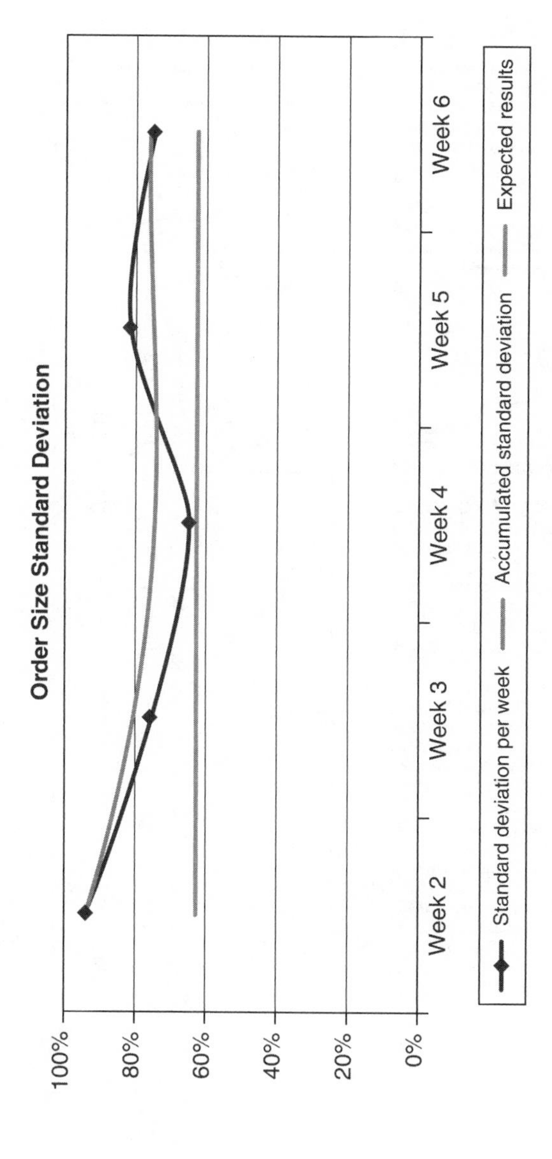

Figure 3.67 Order size

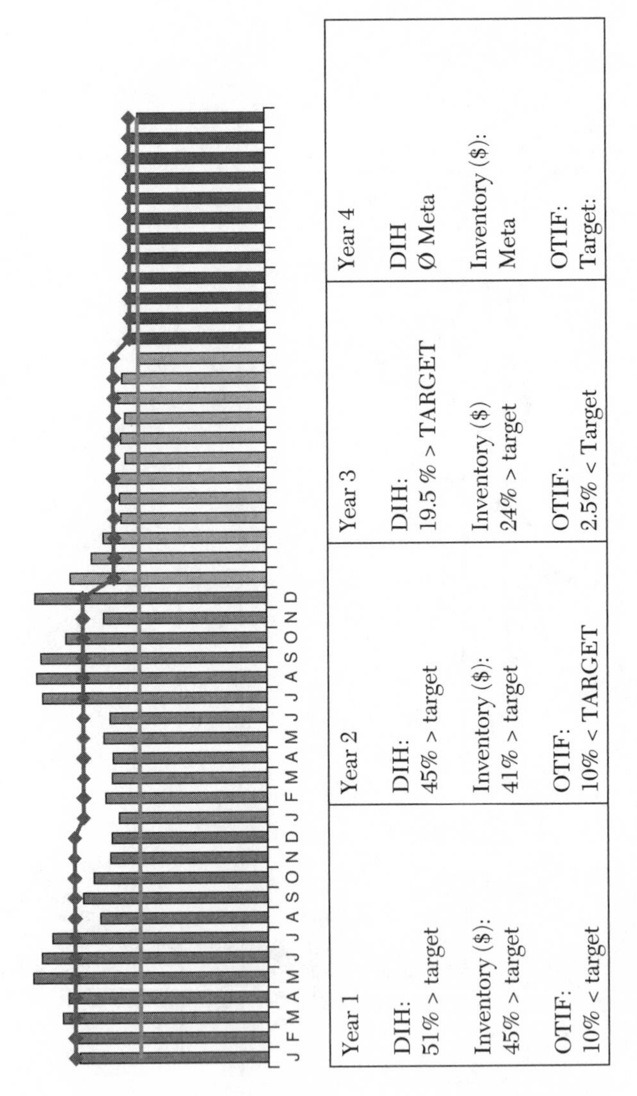

Year 1	Year 2	Year 3	Year 4
DIH: 51% > target	DIH: 45% > target	DIH: 19.5 % > TARGET	DIH Ø Meta
Inventory ($): 45% > target	Inventory ($): 41% > target	Inventory ($) 24% > target	Inventory ($): Meta
OTIF: 10% < target	OTIF: 10% < TARGET	OTIF: 2.5% < Target	OTIF: Target:

Figure 3.68 DIH, OTIF

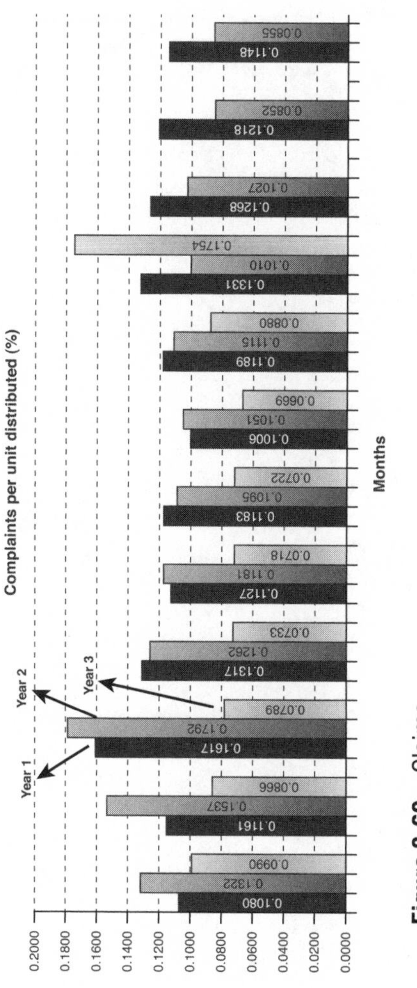

Figure 3.69 Claims

Perfect order: This calculates the orders free of any errors in various stages of processing: orders without errors, processing orders shipped complete, orders shipped on time, and shipped without damages and with complete and accurate documentation. This is an important performance indicator; Section 2.1 provides a detailed example of its use.

4

Managing Change

The impact of performance indicators is produced within the organization. There are three basic dimensions for this impact:

1. How to audit current performance.
2. How to benchmark current performance; it is necessary that current performance has been understood prior to dedicating any effort to benchmarking. The premise to benchmarking is knowing you own the business.
3. How to manage change to meet better performance; this is definitely the trickiest dimension. In this chapter, a business case illustrates several models that explain the impact new performance indicators generate in various levels of the organization.

4.1 Auditing Performance Indicators

To assess the quality of the performance indicator applied to a particular process of a company, we propose a quick self-assessment, answering the following questions with yes or no.

Performance Indicators: Audit Questionnaire		Yes/No
1	The performance indicators in use were based on an initial mapping of the processes of your company, involving the different departments, clients, and suppliers?	
2	Your company has identified the critical processes, among all those listed in the initial mapping?	

Performance Indicators: Audit Questionnaire		Yes/No
3	Was an analysis of interdependence between the critical processes and its input variables conducted?	
4	A methodology was followed to prioritize processes and variations, in order to guide an implementation schedule of performance indicators over time?	
5	To define the performance indicators, an analysis was conducted of the impact of using different input variables in your compiling?	
6	Responsibility for the results of the performance indicators were actually assigned to people who deal with the processes and input variables of these processes?	
7	There is a methodology to promote co-responsibility on the results of a performance indicator? This shared responsibility is built on the relationship of colleagues from the same area or part of the business for an interdepartmental relationship or even involving suppliers, customers, and service providers?	
8	Your company promotes periodic reviews of performance indicators, to evaluate their importance?	

If the answer was no to at least one of these questions, it is possible that your company can review, improve, or just simplify the current portfolio of metrics. Both for companies seeking to implement their metrics, as well as for those who already have them, they are essential to the implementation of a methodology for auditing performance indicators.

These questions make up the pillars of a consistent methodology for companies seeking a continuing increase in the quality of their results by monitoring via a set of indicators.

4.2 Benchmarking

Benchmarking is always challenging. To compare results with credibility, it is expected that the environments in which these results were generated are very similar. Several discussions in Chapters 1 and 2 have shown that these environments may vary in different firms.

A detailed example is shown in Section 2.1 as three different calculation algorithms generate three different outputs for the OTIF (on-time in-full) performance indicator. Section 1.4 presented some attributes that may influence the performance indicator's output and therefore reduce benchmarking precision:

- Variation in the indicator's calculation algorithm
- Expected process learning curve profile and associated time-phased targets
- Stakeholder's requirement related to either report format or frequency
- Organizational structure and impact on managing the control (from data-collection effort to reporting)

Traditional approaches for benchmarking may be categorized into four groups:

1. Benchmarking similar operations within the same business
2. Benchmarking different operations within the same business
3. Benchmarking within competitors
4. Benchmarking processes

A given global beverage industry supports distribution operations in southeast Brazil through a network of regional depots. These depots have a similar physical structure, operate similar volumes, and hold the same product range (SKU[1] portfolio).

[1] SKU = Stock keeping unit

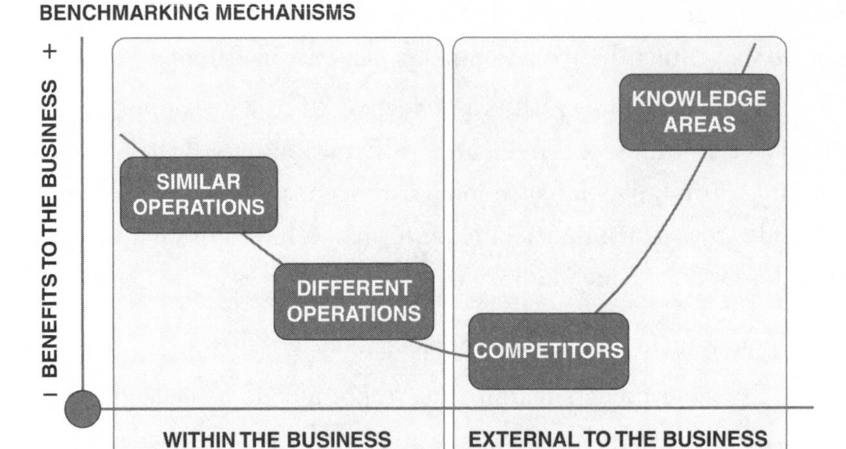

Figure 4.1 Benchmarking mechanisms

This company implemented a performance indicator to quantify the return index of the product shipped from each distribution route. If a truck is loaded with 600 cases and 15 are not delivered, due to several possible reasons, the return index is $(15 / 600) \times 100 = 2.50\%$. This organization established a target for the return index of 1.5%, reported monthly. The result last month is illustrated in the next figure.

This situation allows a detailed process-based benchmarking. By understanding how activities are executed at the best performing depots, some learning may be reapplied to other operations.

However, benchmarking within the same organization is not always possible. Processes often differ significantly, even in apparently similar operations.

A household appliances industry produces refrigerators globally. This company controls the use of a raw material required to manufacture a foam that becomes the equipment's internal protection.

	Depot 1	Depot 2	Depot 3	Depot 4	Depot 5	Depot 6	Depot 7	Depot 8	Depot 9
Return Index	0.9%	1.3%	2.1%	1.2%	0.5%	1.7%	1.4%	1.5%	1.1%
Target	1.5%	1.5%	1.5%	1.5%	1.5%	1.5%	1.5%	1.5%	1.5%
Gap	0.0%	0.0%	0.6%	0.0%	0.0%	0.2%	0.0%	0.0%	0.0%

■ Return Index

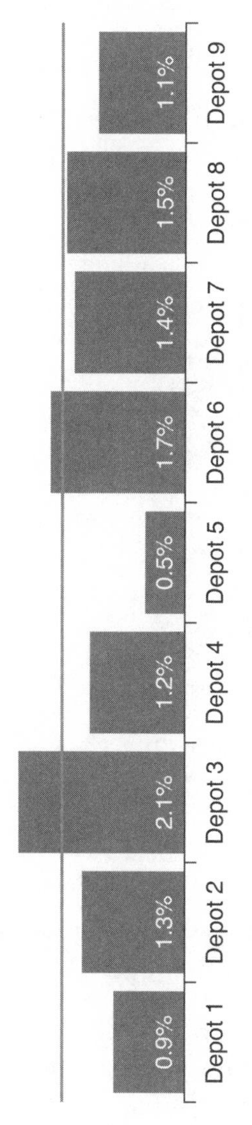

Figure 4.2 Benchmarking (1)

The corporate quality assurance department in Germany defined a standard consumption rate of 0.5 liters per unit produced, which is reflected in the product structure (bill of materials). Two manufacturing sites located in Germany and in Brazil reported the following scrapping index for this item.

	Germany	Brazil
Scrapping Index	1.00%	6.00%
Target	1.30%	1.30%
Gap	0.00%	4.70%

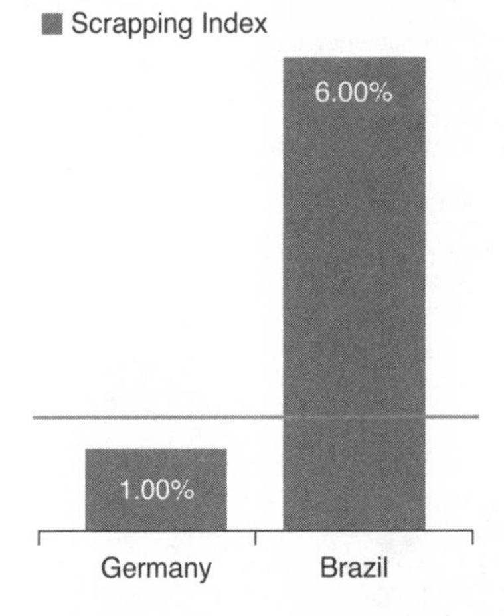

Figure 4.3 Benchmarking (2)

This indicator suggests the performance of the German operation is better than the Brazilian one. But benchmarking these operations would not be possible due to a physical property of this product: volatility. Both manufacturing sites operate at an uncontrolled room temperature.

The average temperature inside the German manufacturing site over the year is 17°C, and inside the Brazilian site, it is 26°C. Given the product controlled by this performance indicator is highly volatile, there is a natural loss that is higher in Brazil than in Germany.

After a few years reporting the need to either review the expected scrapping target for this product in Brazil or investing in temperature control mechanisms, this industry opted for the first alternative, which was the cheaper at the time of the decision. Once they opted to define different targets for the different operations, their benchmarking became less effective.

To recover the benchmarking efficiency, this company decided to report the scrapping index as the percentage of the real scrapping index in each manufacturing site in relation to the specific target defined for each site. Therefore, the expected performance is now 100% (delivering the target).

This example shows how careful the organization needs to be when benchmarking its own processes. The risk is mostly associated with the lack of visibility of the differences within apparently similar processes and using similar indicators to compare diverse environments.

	Germany	Brazil
Scrapping Index	1.00%	6.00%
Target	1.30%	5.00%
Index vs. Target	130.0%	83.3%

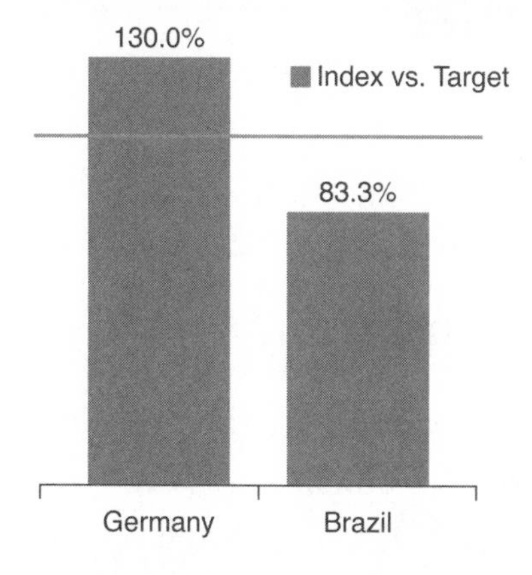

Figure 4.4 Adjusted benchmarking

The next benchmarking category looks at the competitors. This has very limited benefit to the business as it usually ignores the specifics of the environment that generates the indicator. When one company looks at a competitor's performance indicator, some questions are never fully answered, such as the following:

1. What is the scope of the process expected to be monitored by the control?
2. What business premises were considered to define the controlled process?
3. What is the performance indicator calculation algorithm?
4. Is the indicator the result of the consolidation of the performance of subprocesses? If so, how does the mechanism occur?
5. How have the stakeholders influenced the dynamics of this indicator?

A tactic to reduce these uncertainties is to invite sector institutes to lead the benchmarking. Several industries have these organized structures (for example, pharmaceutical, apparel, retail, and banking).

The third benchmarking category is potentially very powerful. The benchmarking of processes sees beyond the company's sector and looks for insights. Once conducted correctly, this approach strongly contributes as a mechanism of organizational knowledge management.

4.3 The Organizational Impact

The number of change management models that offer insightful explanations for organizational movement is proportional to the complexity of the challenges faced by companies of all sizes and vorigins. At the beginning of the first chapter, it is stated that performance indicators should promote the right behavior and the appropriate organizational culture; therefore, it is desirable for a multifunctional approach to define them.

Following internal audits and benchmarking efforts, it is time to implement the selected set of metrics. Although this can be an easy process, most of the time its impact creates tremendous energy within the organization with a serious impact on different organizational levels. The following example is based on a real business case and illustrates the human dynamics through several reference models brilliantly created by Johnson,[2] Conger,[3] Grundy,[4] Blake & Mouton,[5] Morgan,[6] Barach & Eckardt,[7] Stacey,[8,9] House,[10] Blanchard,[11] Johnson,[12] Fisher,[13] Mintzberg,[14] Hersey,[15] and Clark.[16]

[2] Gerry Johnson's Cultural Web Model. Gerry Johnson and Kevan Scholes, *Exploring Corporate Strategy*, 5th Edition, 1999.

[3] Source of failed vision. Conger, J. (1990). The Dark Side of Leadership. *Organization Dynamics, 19*(2), 44–55.

[4] Tony Grundy's Stakeholder Analysis. Strategy implementation and project management. Original Research Article. *International Journal of Project Management*, 16(1), February 1998, 43–50.

[5] Blake & Moutons's Managerial Grid. Blake, R. R. and J. S. Mouton (1964). The managerial grid. Houston, TX: Gulf.

[6] Source of power in organizations. Additional reading: Morgan, Gareth (1993), Images of Organization (London: Sage Publications), ISBN 0-8039-2830-0.

[7] Paradoxes of leadership. Barach, J.A. & Eckhart, D.R. 1998. "The paradoxes of leadership," in Leading Organizations, London: Sage, p 69.

[8] Uncertainty and Agreement, Stacey, R. & Zimmerman, B. (2001). Ralph Stacey's Agreement & Certainty Matrix.

[9] Ralph D. Stacey, Financial Times Pitman Publishing – *Strategic Management & Organisational Dynamics,* 2nd Edition, 1993.

[10] Path-Goal Theory – Leader Behaviour. House, R.J. (1971). A path-goal theory of leader effectiveness. Administrative Science Leadership Review, 16, 321–339. Retrieved on November 28, 2005. Woolard, D. (2005). Path-Goal Theory of Leadership, http://www.drwoolard.com/miscellaneous/path_goal_theory.htm.

[11] Reference model: Situational leadership. *Leadership and the One Minute Manager: Increasing Effectiveness Through Situational Leadership* (with Patricia Zigarmi and Drea Zigarmi, HarperCollins Business, 1985).

[12] Locus of Control and Leadership. The Role of Locus of Control in Leader Influence Behavior, Avis L. Johnson, Fred Luthans, Harry W. Hennessey, *Personnel Psychology*, Volume 37, Issue 1, Pages 61 to 75, March 1984.

[13] The Transition Curve. Fisher, J.M. (2005), A Time for change, Human Resource Development International, vol 8:2 (2005), pp 257 – 264, Taylor & Francis.

[14] Mintzberg, H. (1989). *Mintzberg on Management: Inside Our Strange World of Organizations*, New York: Free Press.

[15] Hersey, P. and Blanchard, K. H. (1969). *Management of Organizational Behavior – Utilizing Human Resources*. New Jersey/Prentice Hall.

[16] Power to the people. G. Clark, *Management Focus,* Issue 12, Summer 1999 (Cranfield School of Management).

Business Case

Change Management and Leadership as Operations Management Tools

Author: Alexandre Oliveira

Part I: First Assignment

FFC is a global fast-moving consumer goods company with a massive global presence. The Brazilian market is still a challenge for FFC since the sales pattern for 2008 is not even close to breakeven level.

The need to reduce supply chain costs and improve service levels through better inventory availability in the newly acquired Healthcare Business Unit in Europe led FFC to decide on a consolidation process, in which all three manufacturing sites (Barcelona, Spain; Toulouse, France; and Milton Keynes, United Kingdom) will be integrated into a major operational facility integrated to a national distribution center close to Bordeaux.

When complete integration is not achieved, FFC has to reverse negative business results of French Plant. This site will be operating for another 18 months and unless a major change is implemented, it may seriously affect the overall results in Europe.

The Toulouse plant employs 300 people and its recent results are not very different from those obtained in the previous quarters and years. Low productivity impacts on product cost and low sales worsen the situation, while product quality indexes are among the worst within similar sites of the globe.

This case presents the organizational effort made to sustain and leverage operational standards and employees motivational levels within the 18 months until complete shutdown.

The Cultural Web Model was developed by Gerry Johnson and Kevan Scholes and presents six dimensions that define a so-called paradigm (the pattern or model) of the work environment: stories, rituals and routines, symbols, organizational structure, control systems, and power structures.

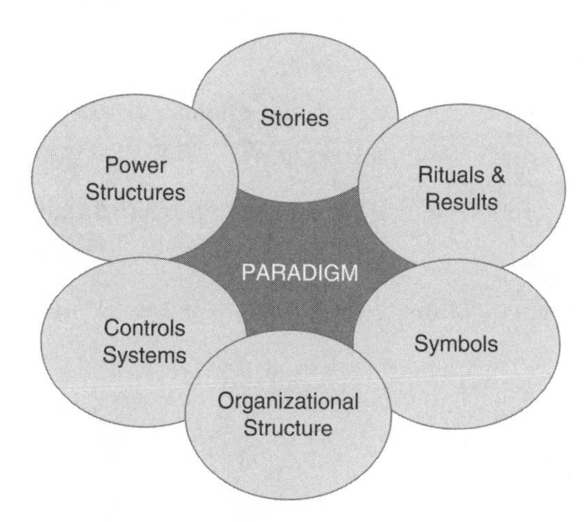

Figure 4.5 Cultural web

The Cultural Web for the Toulouse Plant

Routines: All managers are supposed to arrive at 8 a.m. All managers attend a daily meeting at 8:30 a.m. After the 8:30 meeting, managers used to have breakfast in a bar in front of the factory. Usually, managers do not start working before 9 a.m. On Thursdays, managers have their weekly happy hour. On Fridays, managers, operators, technicians, and staff meet to play football.

Rituals: Playing video game in the shopping center with the plant manager; business results review, goals deployment; corporate training; feedback sessions.

Symbols: Managers at level 3 have a company car; only managers can park inside the manufacturing site. Only senior managers and the plant manager have their offices on the second floor. Operators and technicians struggle to have a PC; it is a symbol of power.

Stories: FFC is a company that pays very well. FFC is extremely demanding, especially in relation to the managerial level. FFC hires the best people. FFC provides internally the tools for professional growth (no need for external tools). FFC provides a quick career growth. Career in production is as quick as in finance or marketing. FFC provides a job for life.

Control systems: Business review, annual quality assurance audit, benchmark of all measures, feedback sessions.

Power structure: Production managers are more powerful than other managers at the same level.

Organizational structure: Managerial structure is based on a vertical approach.

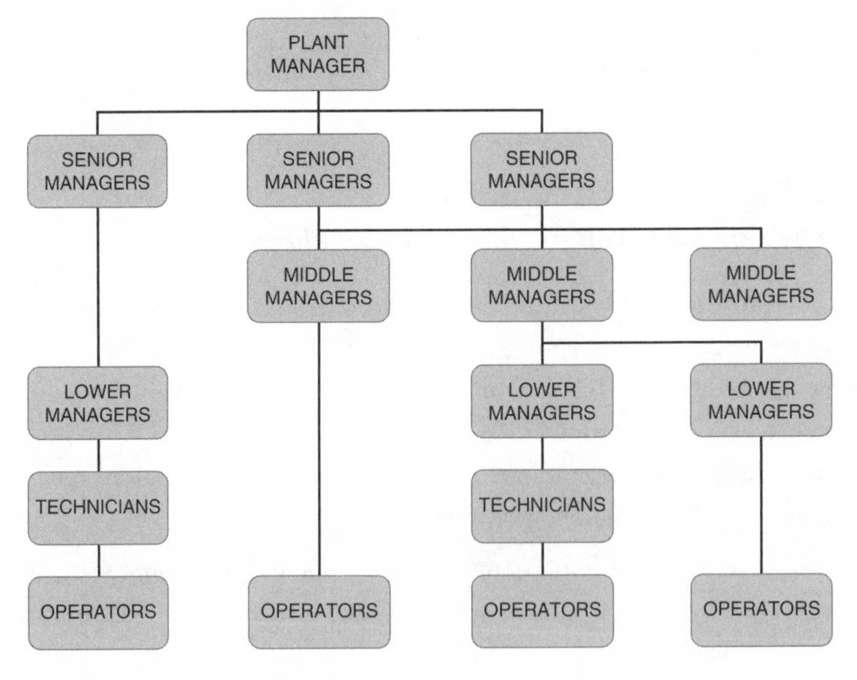

Figure 4.6 Organizational structure

Paradigm: FFC develops and makes premium products of high quality to provide its consumers with the best value added products available in the market

Conger in his model Source of Failed Vision suggests four causes for organizational bad governance:

1. Vision reflects the internal needs of the leader, not the market:

   FFC is a visionary company, and its global vision must be deployed to all organizational structures and levels, including the manufacturing sites. During this long deployment process, eventual distortion may affect the original vision. The manufacturing sites have several measurements and indexes to track whether operational results are synchronized with established goals. But the Toulouse plant barely manages to track the results properly toward the vision, for both external and internal reasons.

2. Resources requirements have been seriously miscalculated:

   Some reasons are external to the plant scenario, such as FFC's failure to recognize environmental changes, resulting in a distorted view of market needs. As a consequence, the sales level is far below the expected, impacting directly on the operational costs.

3. Distorted view of market needs:

   FFC continues to develop the Brazilian market for premium products. but its leading brand is more expensive than that of its competitors and its market share is insignificant. This scenario does not create consumer added value to its products and therefore affects sales Figures.

4. Failure to recognize environmental changes:

   Internally, the major reason for not properly conducting the business toward the company's vision is the misleading, miscalculation of resources, both in terms of quantity and quality. In addition, most marketing initiatives seem to reflect personal assumptions and beliefs rather than real consumer needs.

The plant manager, who joined the Toulouse plant as junior production manager (level 1), had assignments in inventory management and the quality assurance department before assuming the current position. Most of his current subordinates were his peers until very recently and share common personal interests, which has led to conflicts of interest.

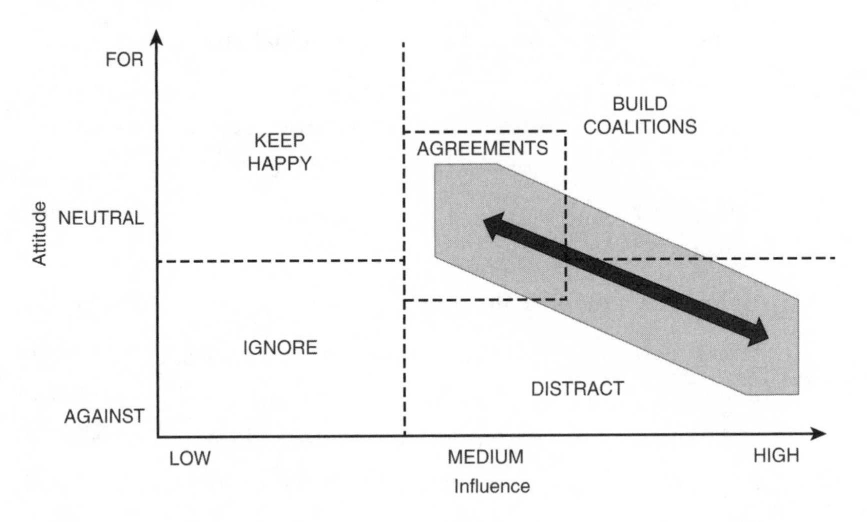

Figure 4.7 Stakeholder analysis, Tony Grundy

Although the plant manager has the authority to make powerful decisions, he rarely uses his power in a positive way (to create value for the business). As a consequence, senior managers (direct subordinates) are always oscillating between agreement and distraction. This distraction enables a fuzzy environment that leads the plant to produce poor results.

The Toulouse plant is a small manufacturing site, with 275 operators and technicians and 25 managers (levels 1 to 3). The managers resemble a society where the president is the plant manager. The plant manager often invites his team to have lunch in shopping centers where they can play video games. The plant manager needs harmony and is good at creating it.

The *Managerial Grid* Reference Model clearly illustrates the imbalance between the focus on people satisfaction and the focus on needed business results.

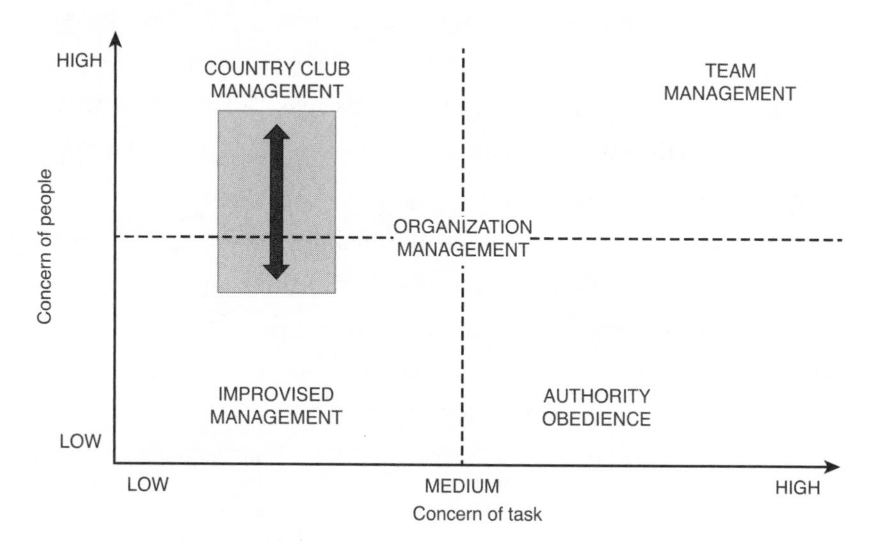

Figure 4.8 Managerial Grid

According to Morgan, the sources of power in organizations include formal authority, control of scarce resources, use of organizational structure, rules, regulations, control of decision processes (and premises), control of knowledge and information, control of boundaries, ability to cope with uncertainty, control of technology, alliances and network, control of counter-organizations (unions and lobby groups), symbolism and management of meaning, the management of gender relationships, and define structural factors.

Morgan's model for sources of power in organizations helps us to understand some reasons why the plant manager fails to attain power. The plant manager has a limited ability to cope with uncertainty and does not have total control of the decision process; he is always open to a new, final option. Although it is clear he has formal authority, he uses it to delegate most of the decisions related to resource allocation and cannot cope with scarce resources. Therefore, the plant's overheads have been kept within high figures for many months.

Barach and Eckhardt list the following paradoxes of leadership: (1) autonomy versus dependence; (2) empowering others versus power to command; (3) formal authority versus granted by the governed; (3) archetype of group norms versus agent for changing them; (4) share information to gain loyalty versus keep the power of knowledge; and (5) role versus person.

The lack of a clear vision and the inability to set tangible goals pushed the organization beyond the point of making rational decisions or even making judgmental or ideological decisions. An anarchy (or avoidance) scenario is not achieved but only due to a strong effort (mainly by the plant manager) to make people come to ultimate agreements. Stacey's Agreement & Certainty Matrix illustrates this behavior.

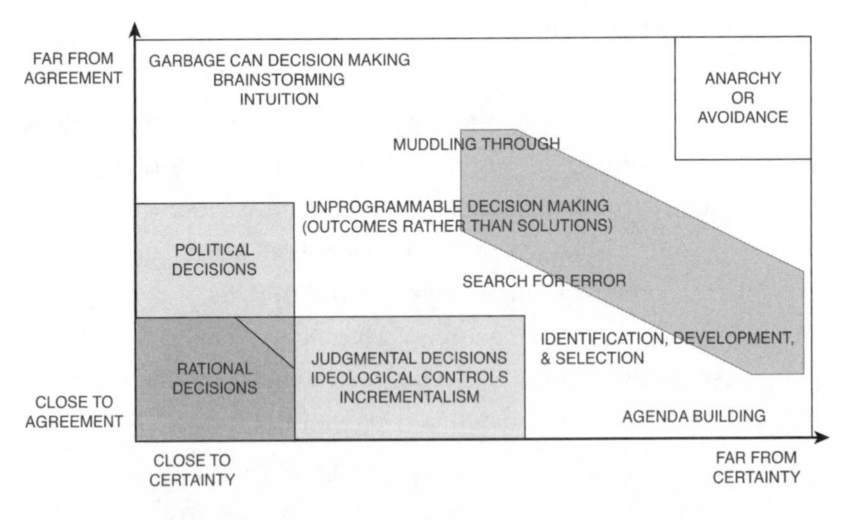

Figure 4.9 Agreement & Certainty Matrix

The path-goal theory states four behaviors typical of the leaders:

1. Directive leadership
2. Supportive leadership

3. Participative leadership
4. Achievement-oriented leadership

In the Toulouse plant, there was a clear participative leadership. Blanchard's situational leadership model illustrates that.

As a recent organization, the Toulouse plant is full of new hires. FFC's recruitment policy is based on only recruiting recent graduates from top universities. Therefore, the average working experience for middle managers is 5 years since graduation, and lower managers have an average of 1 year of working experience.

Lower (level 1) managers are 65% of the management level. There is a clear expectation from the company that even the new hires can cope with senior tasks. The model for situational leadership (Figure 4.10), illustrates from the point of view of the company (and the plant manager's), that the team should be empowered to act independently with appropriate resources to get the job done (dark circle on the right).

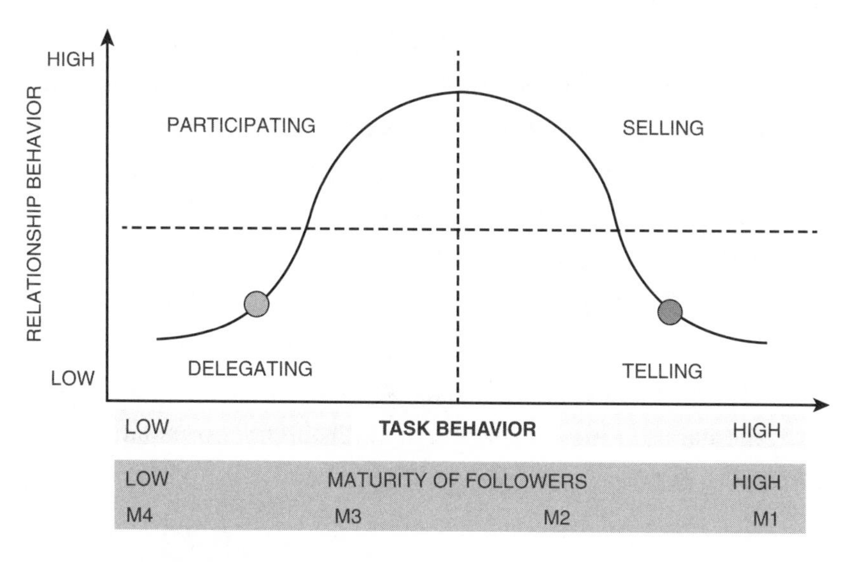

Figure 4.10 Situational leadership

But the young managerial team is in need of either a delegating or participative approach (light circle on the left). Also needed is a supportive approach, where the leader and the employee can make decisions together. In this case, the role of the leader is to facilitate, listen, draw out, encourage, and support (not just tell).

Nevertheless, not only are these resources miscalculated (as mentioned previously) but also the team is very young. Not all members of the management team are ready for the tasks and most of them are not psychologically ready. This effect is discussed ahead.

Hersey and Blanchard suggest three leadership skills:

1. **Diagnosing:** Understanding the gap between the current situation and what can be expected (the performance gap).
2. **Adapting:** Changing one's behavior and other resources to close the performance gap.
3. **Communicating:** Enabling others to understand and accept what must be done.

For many quarters (almost 2 years), the Toulouse plant has not achieved the expected results. Senior managers have not been able to understand the gaps between the current situation and what is expected (performance gap). This scenario is completed by a failure to change current behavior and resources to close the few known gaps. The communication process has not been effective in the same way that the organization did not respond to the need for change.

According to the path-goal theory, the leader's behavior is acceptable if (1) it is seen as an immediate source of satisfaction or directly instrumental in future satisfaction or (2) if the followers' satisfaction is linked to their own perception of their abilities (locus of control).

According to the Locus of Control and Leadership Model, at the Toulouse plant, immediate subordinates are comfortable with their reporting level. Because FFC is still building its structure in Brazil, positions in upper management levels are often created, and the current plant manager closely follows his subordinates' careers. At the same time, his exceedingly participative managerial style does not pressure the subordinates with tight due dates or with overwork.

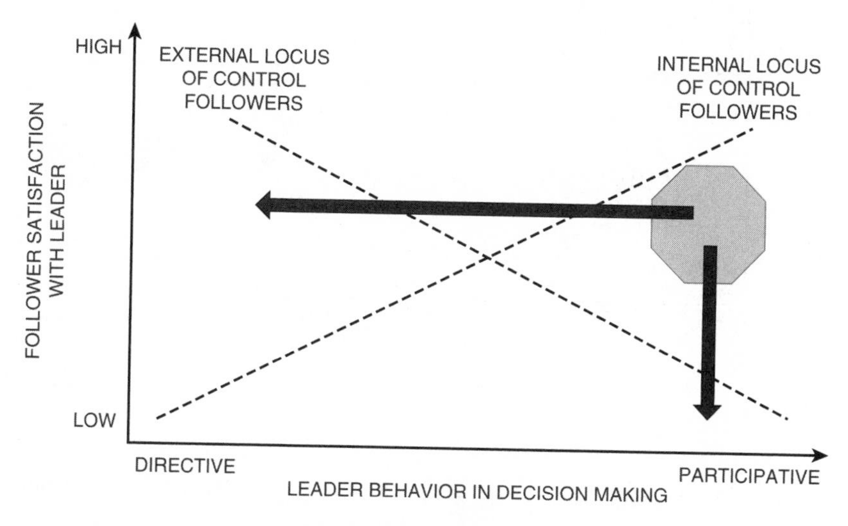

Figure 4.11 Locus of control and leadership

The reference model for psychological moment, illustrated in the next figure, enables us to understand how different levels of workers accommodate the trade-off between perceived discretion and their own empowerment. FFC used to hire high-level technicians who were subordinated to lower managers (recent graduates). This was a frustrating situation for the technicians because they expected more freedom to develop their abilities, and it also caused anxiety for the lower managers' group.

Lower (young) managers are expected to lead and coach experienced technicians which means to plan their careers including future promotions and salary reviews. Lower managers are also expected to acquire technical expertise within a minimum period of time to cope with troubleshooting analysis and some decisions such as hiring temporary workers or preparing a budget for the next fiscal year.

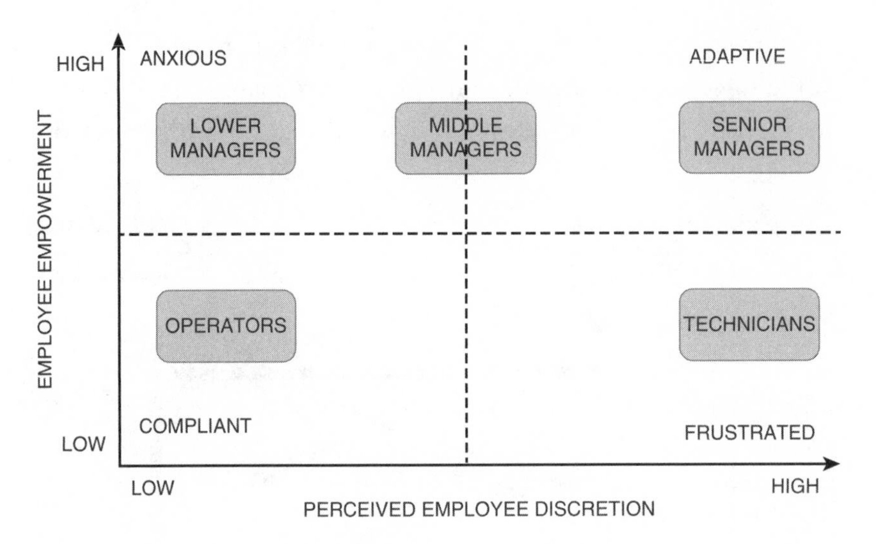

Figure 4.12 Psychological moment

As the manager become more experienced, he learns how to deal with high expectations, and middle managers tend to cross the border to the adaptive corner. Senior managers are already in the adaptive corner.

Part I: Second Assignment

A few weeks after the last results review, it was announced that the current plant manager received a new assignment as special projects director. Within a month of this announcement, the former Toulouse plant manager left FFC. For 5 weeks, four operations managers ran the Toulouse plant as a collegiate, while the name of the next plant manager was not announced. In January of the next year when the name of the new plant manager was finally announced, most middle and senior managers had a very negative reaction.

FFC was in the middle of a consolidation process. When FFC entered the European market, it bought three manufacturing sites in Toulouse, France; Milton Keynes, United Kingdom; and Barcelona, Spain.

A few years later, new land was bought at Bordeaux, France, a city 2 hours by car from Toulouse. The strategy was to consolidate the three plants into a single site and build a national distribution center. The operations shutdown of both Milton Keynes and Barcelona occurred before the Toulouse plant changed its plant manager.

The plant manager of the English plant, after completing the transition to Bordeaux, was assigned as the new Toulouse plant manager. Some managers at the Milton Keynes plant had been transferred to the Toulouse plant and knew their previous leader's style. Some managers were transferred from Milton Keynes to the new site in Bordeaux.

The new Toulouse plant manager is known to have an extremely directive approach to business decisions, which is the opposite style of his predecessor. As the new plant manager exercises his power, all other management levels may lose power.

The immediate consequence is illustrated in the following. Lower managers, inexperienced professionals, moved from the anxious corner to the compliant corner. This is, from the point of view of the lower managers, a positive action. Basically, lower managers felt more comfortable with this situation once they understood the opportunity to gain some time (and know-how) before more important business expectations relied on them.

The worst impact of the power concentration on the new plant manager happened to middle managers who moved from the adaptive corner to the anxious corner. They are professionals with a working experience of 3 to 5 years and the psychological changes impacted on them more than their job routine.

Several final decisions about the key aspects of daily management were absorbed by the senior managers. They were not in the frustrated corner because, despite the directive behavior of their new leader, he (the plant manager) was known to be a winner in his career and was also known to bring people up with him. Therefore, despite general unhappiness due to the loss of power, middle managers decided that in the future they could benefit from this situation, which brings us back to the path-goal theory.

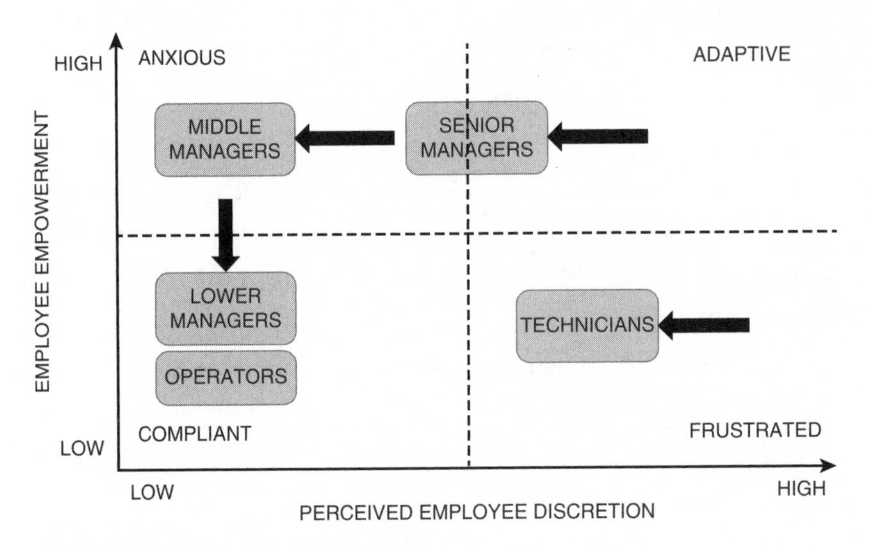

Figure 4.13 Psychological moment, 2

Senior managers, as more experienced professionals, knew how to deal with this situation. Although the loss of power at their level was extensive, the ability to play political games helped them to keep some psychological and power stability. Most senior managers intended to play the sponsorship game, where they could hitch themselves to the plant manager.

But definitely the most effective game was the budgeting, where the four senior operation managers made huge efforts to gain control over resources. The budgeting game was once very dynamic, due to the seasonality of the different products; the power floated from one operational manager to another.

Although different employees found an easier or harder way to adapt to the new imposed situation, most of them were clearly dissatisfied with this situation. The following figure shows how they moved from a more comfortable relationship with the former plant manager to a delicate one.

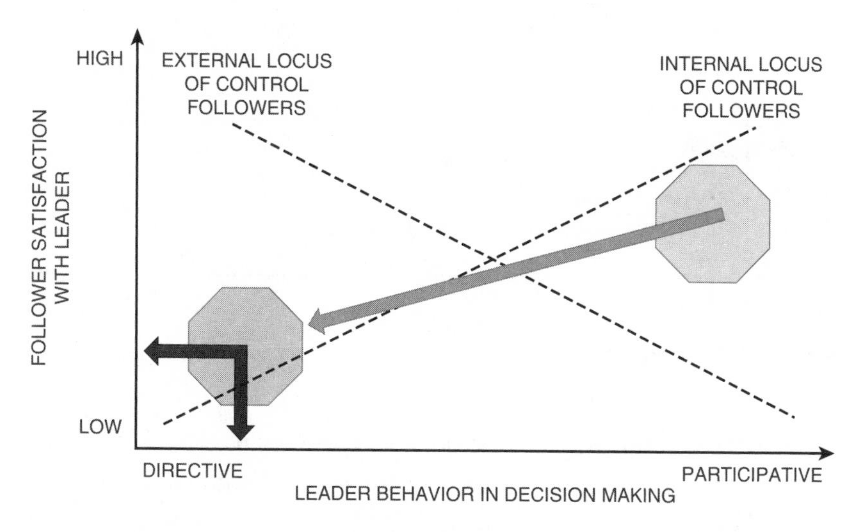

Figure 4.14 Locus of control and leadership

The reference model of the transition curve illustrates, in the next figure, the moment of this organization. The first step is the shock, when the name of the new plant manager was announced. This first step extended for the initial weeks of the new plant manager's administration. In this period, most of the questions he asked could not be responded to because there was no system or measurement tool implemented to provide accurate answers. The following few weeks marked the second step on the transition curve. Most managers, especially the middle ones, insisted on denying the need for change.

As soon as the new plant manager noticed some deep organizational distortions related to task delegation, job descriptions, and others, he worked hard to correct this route. There was not a single basic approach for all managers, and the plant manager tried to develop specific models for each management level.

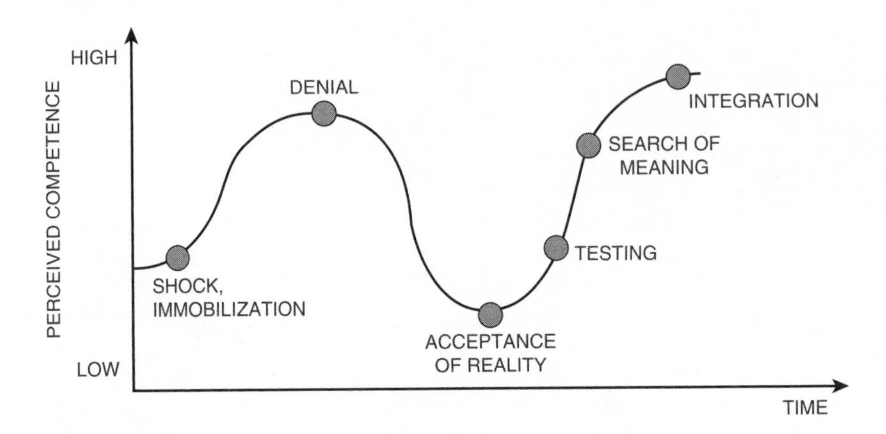

Figure 4.15 Transition curve

The first thing was to relocate the lower managers. They were repositioned in such a way that the organization should focus on training them rather delegating tasks.

These managers at level 1 had an initial period more focused on the Telling corner, but at the same time they were in the Coaching corner. After several months, some of them would move to the Participating corner; the speed of this cycle only depended on individual performance.

Although the middle managers, more experienced than the lower ones, remained in the Delegating corner, they received a more supportive approach from the plant manager. Some of the middle managers could then move into the Participating corner, according to individual performance and maturity. They were also responsible for some key business decisions. Nevertheless, due to the directive leadership, they were often told what to do, and this recurrent presence in the Telling corner frustrated most of them.

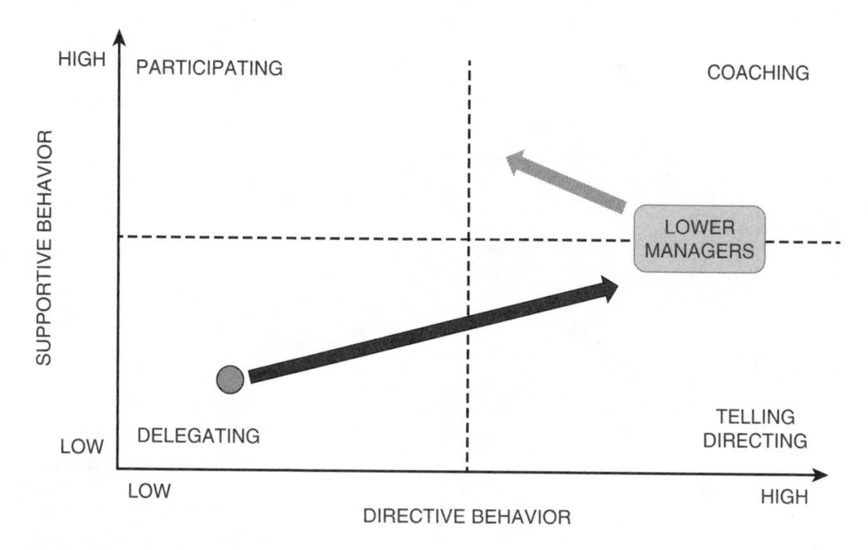

Figure 4.16 Situational leadership, 3

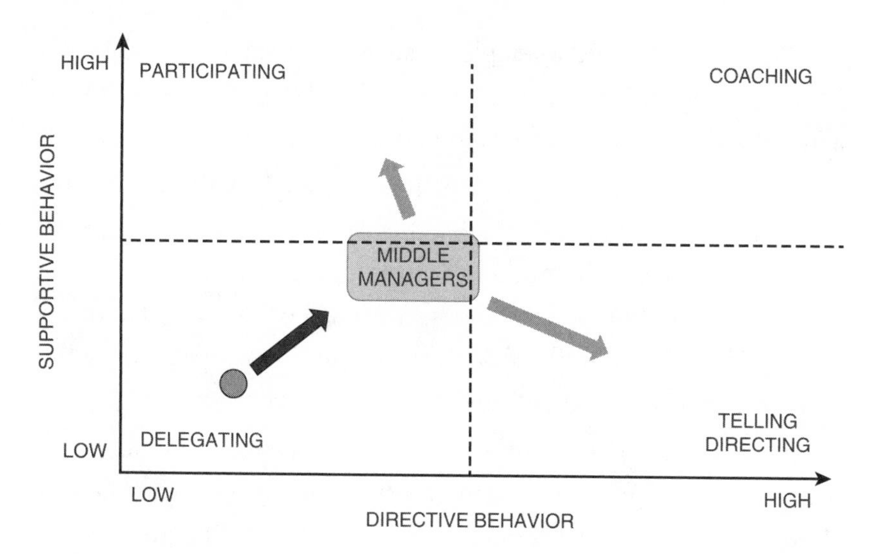

Figure 4.17 Situational leadership, 4

Finally, the senior managers were repositioned. A reasonably high participative and decision-taking freedom was dispensed to them (when compared to the other managers), but, they still had to cope with the directive behavior of their superior, which kept them partly in the Directing corner.

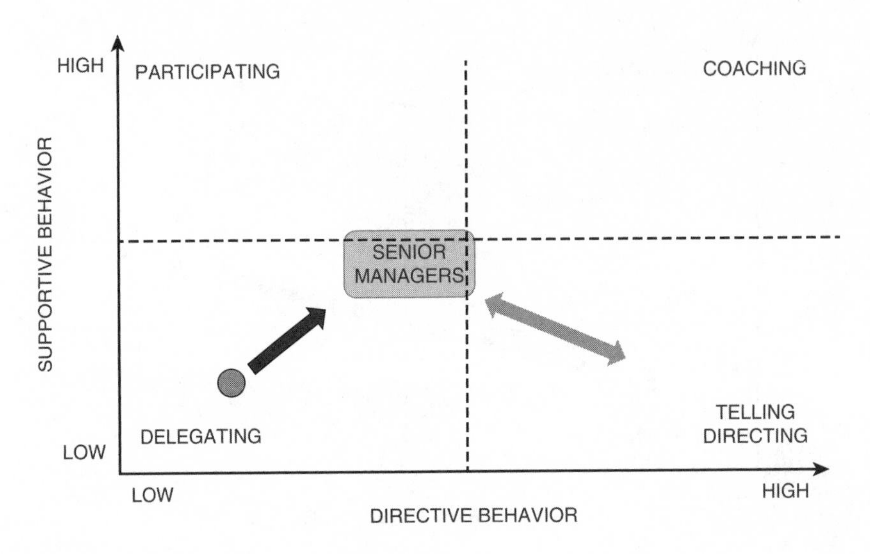

Figure 4.18 Situational leadership, 5

The relationship between plant manager and the managerial team has also changed. The previous manager had kept this relationship as a low concern for tasks but a wide range of concern for people.

The new plant manager drastically increased the task concern profile of the organization, but his directive style kept relationships with employees nearer to the Authority/Obedience corner, although many decisions concerning people development had been taken. There is a common belief that, in the future, the organization will move to the Team Management corner; see the following model.

As with the snowball effect, all these changes impacted on all decision-making processes. However, these processes are now much closer to agreement and certainty areas (see the next figure). Nevertheless, this profile relied on the directive leadership rather than on teamworking.

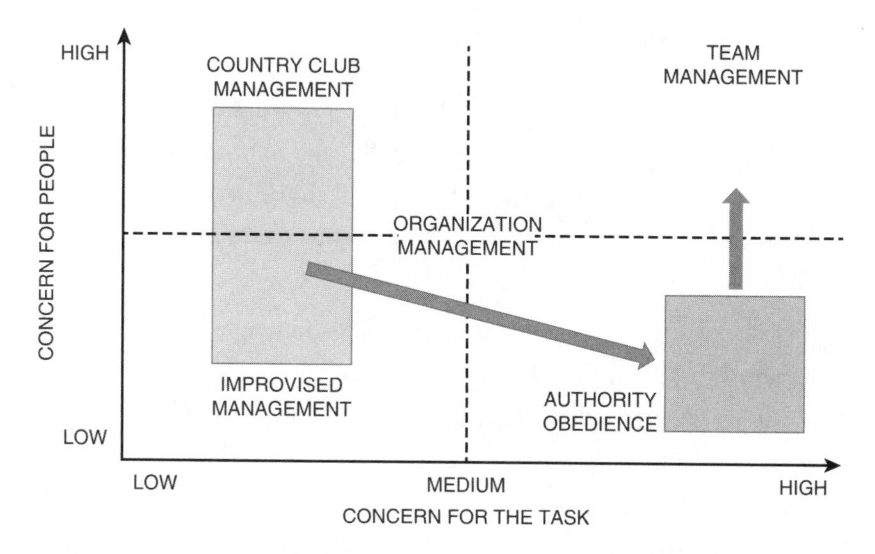

Figure 4.19 Managerial Grid 2

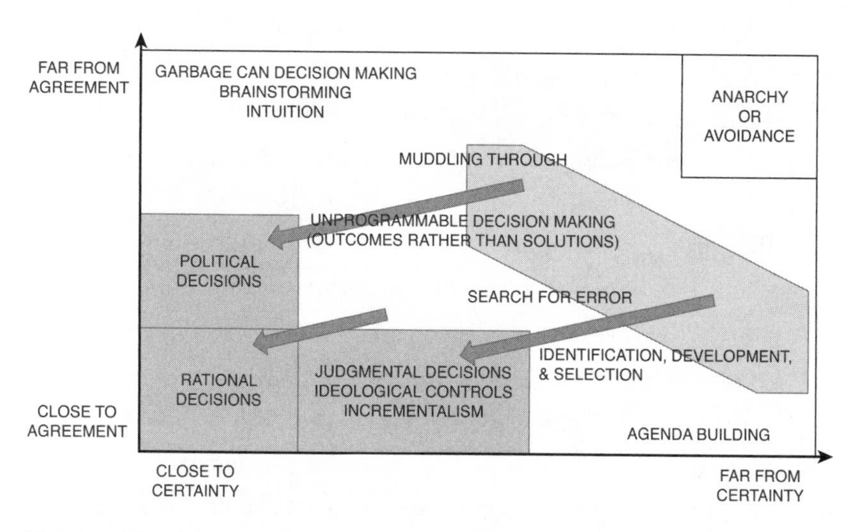

Figure 4.20 Agreement and Certainty Matrix

Returning to the Transition Curve Model, Steps 3 and 4 have been achieved. A few simple and immediate results encouraged most people to follow the new leader.

Part II: The Next 18 Months

After an initial period of organizational realignment, the new plant manager opted for a clear strategy to run the business. Although it was known that the Toulouse plant would shut down within several months, he aimed to have a continuous improvement environment until the last day of operation.

He implemented a housekeeping philosophy and many areas in the plant were painted, tidied up, and even remodeled. His goal was that people would really feel sorry to leave the Toulouse plant. "I want people to miss this place when we are gone," he used to say.

Soon results were achieved. Costs improved, morale was raised, and an effective communication strategy brought representatives from other manufacturing sites to visit the Toulouse plant. Representatives from the Americas, Europe, and Asia came to learn from the implemented systems.

The new plant manager knew how to amplify the achieved results through the company's international channels. All results would be monitored with visual tools and control boards were spread all over the office area and the shop floor. Lower and middle managers used to say that the Toulouse plant was transformed into a new Louvre Museum.

As results became evident, managers' careers benefited and their satisfaction with the leader also increased: the Locus of Control Model (next figure). In this period, as the plant manager trusted in the development of his managers (and as they became more mature), he became slightly less directive.

The organizational environment seemed to be in a good position. More mature managers could cope with higher complexity tasks and the business results benefited. The organization was now working more like a team, although the plant managers directive behavior was always present, as shown in the next figure.

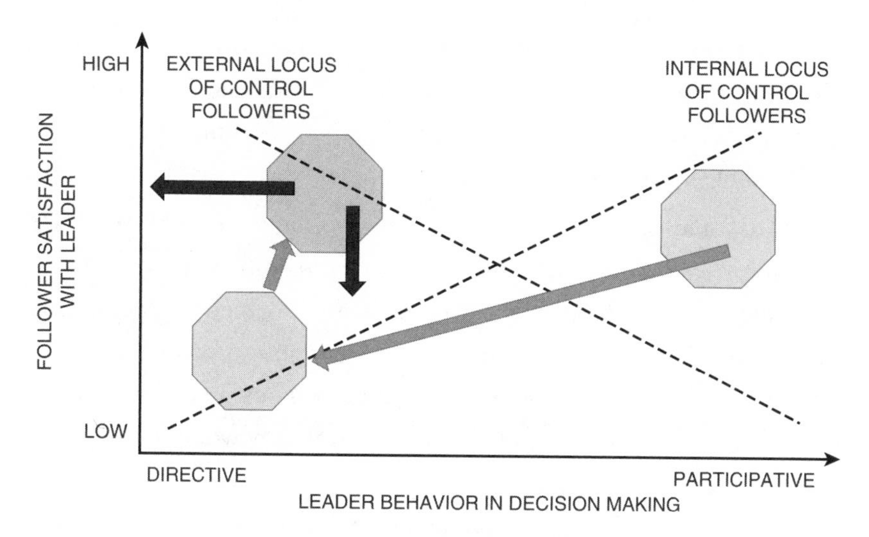

Figure 4.21 Locus of control 2

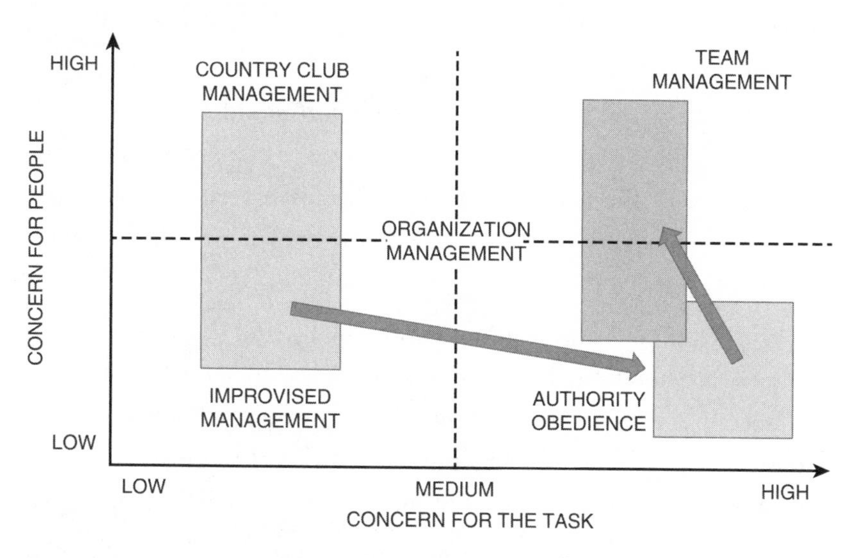

Figure 4.22 Managerial Grid 3

In the model of the transition curve, the point of integration has now been achieved, although some points were not brought to an ideal location.

The tasks led by the new Toulouse plant manager were not easy. He had to retain organizational morale (after raising it) and retain good-quality results (after increasing them) at the same time that he had to shut down operations and fire people.

The Toulouse plant had three production lines (P1, P2, and P3). The first area to shut down was P1. Most of the employees were fired, and a few of them (the best performers) were relocated to other areas. Middle and lower managers had to decide whose employees would stay or would be fired. This communication was never an easy task. In one-to-one talks, employees were informed if they would be kept on in the company.

The same process occurred when P2 operation was shut down. Although most people had the option to choose between staying in the company or moving to the Bordeaux plant, the final decision was always painful.

P3 operations ran alone for 1 month before shutdown. This month was terrible for the morale of the group. Lower managers and operators felt heavily the consecutive dismissal process. Although middle and senior managers were adapted, the overall morale of the plant was heavily impacted.

When the Toulouse plant operation finally shut down, the overall feeling was that although the organization had crossed a hard time, it had become more mature and that the whole process actually made a positive contribution to the business and to the employees that remained at FFC (both in terms of career and personal development).

If the real strategy of the plant manager was to make people miss their plant when it was gone, he achieved it. A lot of work and effort was put into making the Toulouse plant a better place in terms of performance. To those who decided to stay within FFC and go to the Bordeaux plant, there was a real feeling of success.

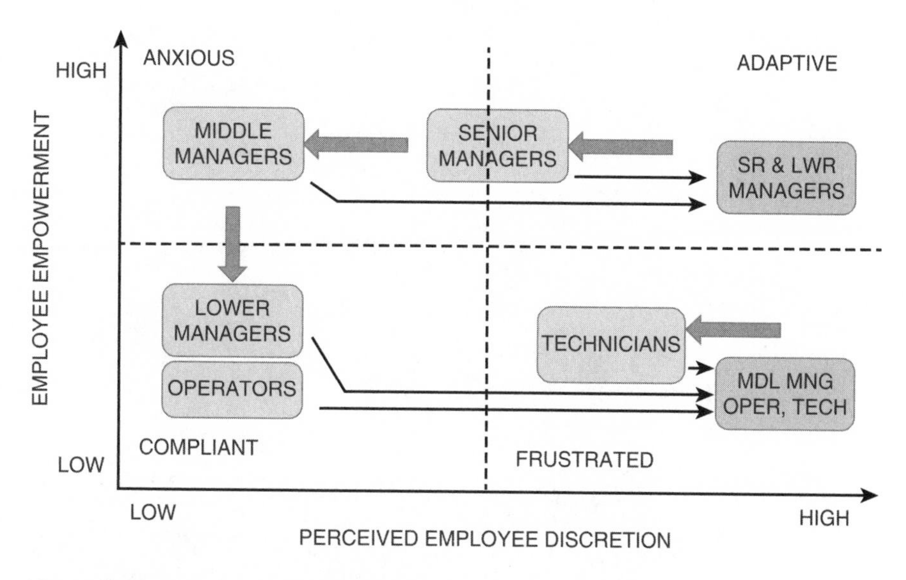

Figure 4.23 Psychological moment 3

Only a few of the original 300 employees of Toulouse plant went to Bordeaux. More than 220 employees either decided to leave the company or were fired due to low performance. Some 25 of the 80 employees that moved to Bordeaux were managers. This means that only 55 out of 260 (21%) technicians and operators went through this painful process.

The few technicians and operators who went to Bordeaux kept in mind the memories of their friends being fired. Most of those employees never had the same commitment, as shown in the last months of the Toulouse plant.

Bibliography

Oliveira, A. and Gimeno, A. *A Guide to Supply Chain Management: The Evolution of SCM Models, Strategies, and Practices.* New York: Pearson, 2014.

_____. *Customer Service Supply Chain Management: Models for Achieving Customer Satisfaction, Supply Chain Performance, and Shareholder Value.* New York: Pearson, 2014.

_____. *Managing Supply Chain Networks: Building Competitive Advantage in Fluid and Complex Environments.* New York: Pearson, 2014.

_____. *Supply Chain Management Strategy: Using SCM to Create Greater Corporate Efficiency and Profits.* New York: Pearson, 2014.

Index

D

E

F

G-H

I-J

K

L

M

O

P

T

U

V

W-X-Y-Z